Seafood: Resources, Nutritional Composition and Preservation

Seafood: Resources, Nutritional Composition and Preservation

Kamal Duggal

CYBER TECH PUBLICATIONS
4264/3, Ansari Road, Daryaganj, New Delhi-110002 (India)
Ph.: 011-23244078, 011-43559448 Fax: 011-23280028
E-Mail: cyberpublicationsdelhi@yahoo.com
Website: www.cybertechpublications.com

SEAFOOD: RESOURCES, NUTRITIONAL COMPOSITION AND PRESERVATION

Kamal Duggal

First Edition 2010

, Asian Offset Printers

Published by :

G.S. Rawat for **Cyber Tech Publications**
4264/3, Ansari Road, Daryaganj, New Delhi-110002 (India)
Ph.: 011-23244078, 011-43559448 Fax: 011-23280028
E-Mail: cyberpublicationsdelhi@yahoo.com
Website: www.cybertechpublications.com

Preface

Research and development of seafood continues to be productive in terms of new and improved products for both food and non-food purposes. The use of biotechnology, microbiology, computer modeling and advanced analytical techniques has led to improvements in processing and product safety. This recent book provides extensive new information on these developments.

Aquaculture is a relatively new development in the region, and in most Pacific Islands where it has been attempted; its history goes back less than 30 years. There is no fund of traditional knowledge for culturing fish and shellfish, just catching them, except in very specialized instances and areas. There is thus no great resource of aquacultural skill or infrastructure. This steep development path has perhaps not been taken into account in some development projects, which have often had unrealistic short-term aims and lacked follow-up. Despite the comparatively minor penetration of aquaculture into Pacific Island economies, and despite the loss of interest by most of the international development community after many short-term project failures, several Pacific Island governments have accepted the challenge. They recognize that expansion in capture fisheries is limited, and have made substantial investments in freshwater aquaculture and mariculture, often in concert with external sources of development assistance.

This is especially so in this case due to the lack of effective management of fisheries resources, which, as described, may only add to resource depletion under trade liberalization. Given the way the WTO does not monitor such local dimensions, it is incapable of assuring any outcome outside that which relates to the state as a single economic entity.

—Author

CONTENTS

1

INTRODUCTION

Compared with fishing, aquaculture is currently of little commercial significance to the Pacific Islands, with one important exception, black pearl farming, which is virtually confined to eastern Polynesia. Elsewhere in the Pacific, considerable development is needed before aquaculture can be considered economically sustainable.

Shrimp (Penaeus spp.) farming has been a focus of commercial development in several islands with varying degrees of success; tilapia (Oreochromis niloticus) aquaculture has entered the subsistence economy in some areas, and seaweed (Kappaphycus spp.) is a future commercial export prospect. The culture of other marine and freshwater species is, however, generally still at the experimental or "backyard" stage. The expansion of aquaculture in the Pacific will depend on providing better production methods for species currently being farmed, and techniques for propagating and growing the "new" species described above.

These methods and techniques should be simple and flexible so that they can be adapted to the context of the Pacific Islands environment and to the market constraints (local and export markets). This approach should favour systems integrating fisheries and mariculture with low investment and operating costs and simple technical production processes. This should be done in association with pilot commercial-scale operations to test and demonstrate the economic viability of the methods proposed. This will require research combined with assistance, training and education programmes. Pacific Island nations have many attributes that favour development of aquaculture and stock enhancement in the coastal zone.

These are as follows: a great diversity of coral reef species which are in high demand, proximity to major aquaculture and seafood markets in Asia, availability of suitable growout sites in pristine habitats, geographic conditions which favour restocking and stock enhancement, a relatively inexpensive labour force, and a tradition of working with marine resources. Although these confer many advantages on the region in terms of aquaculture development and stock enhancement, there are still several constraints for such enterprises in the Pacific, which include limited domestic markets, high added-value export markets targeted, transport problems, socio-economic factors, fragile habitats, limited fresh water, and cyclones. Some of the best opportunities for aquaculture development in the Pacific are in the aquarium trade (coral reef fish, hard corals, soft corals), the live seafood markets (e.g. groupers, spiny lobsters, abalone, crabs) and the pharmaceutical industry (e.g. algae, sponges, soft corals). In all cases, the products are of high value and can be grown in small areas with relatively simple technology. This review covers the insular Pacific as defined by the work-area of the Secretariat of the Pacific Community (SPC).

Compared to fishing, aquaculture is currently of little commercial significance in the insular Pacific, with one important exception-black pearl farming-and this is virtually confined to eastern Polynesia. Elsewhere in the Pacific, considerable development is needed before aquaculture can be considered economically sustainable. Shrimp (Penaeus spp.) farming has been a focus of commercial development in several islands over the past 30 years, with varying degrees of success; tilapia (Oreochromis niloticus) aquaculture has entered the subsistence economy in some areas, and seaweed (Kappaphyceus spp.) is considered a future commercial export prospect by the region. The culture of other marine and freshwater species is generally, however, still at the experimental or "backyard" stage.

Aquaculture is a relatively new development in the region, and in most Pacific Islands where it has been attempted; its history goes back less than 30 years. There is no fund of traditional knowledge for culturing fish and shellfish, just catching them, except in very specialized instances and areas. There is thus no great resource of aquacultural skill or infrastructure. This steep development path has perhaps not been taken into account in some development projects, which have often had unrealistic short-term aims and lacked follow-up. Despite the comparatively minor penetration of aquaculture into Pacific Island economies, and despite the loss of interest by most of the international development community after many short-term project failures, several Pacific Island governments have accepted the challenge. They recognize that expansion in capture fisheries is limited, and have made substantial investments in freshwater aquaculture and mariculture, often in concert with external sources of development assistance.

Pacific Islanders cannot turn away from the sea. It is the greatest resource they have.

The Potential Importance of Aquaculture to Pacific Islands: A major problem facing most of the island nations in the Pacific is that they have relatively few opportunities to generate income (Adams, 1998). The economies of most Pacific countries are limited due to small landmasses, few terrestrial resources and low numbers of inhabitants. To ensure further development, island nations must make the most of the one important resource they all have-the sea. Through the joint efforts of the Forum Fisheries Agency (FFA) and the SPC Oceanic Fisheries Programme, island nations are deriving major inputs to their economies by fishing for tuna, or by selling access rights to tuna, within the large maritime zones under their control. However, valuable, sustainable harvests are also possible from inshore waters and coral reefs.

The potential for increased well-being of coastal communities from the responsibleuse of their inshore marine resources arises because the inshore habitats surrounding Pacific nations support a great diversity of economically important species (Wright and Hill, 1993; Dalzell et al., 1996; Bell and Gervis, 1999). Traditionally, these animals were harvested at subsistence levels. More recently, however, development of export markets has provided coastal communities with opportunities to earn income from the inshore fisheries species. Unfortunately, the transition from a subsistence to a market economy has usually been far from ideal: chronic overfishing has occurred in some areas. In such places, there are now too few of the prized animals to sustain reasonable harvests. Destructive fishing methods have compounded the problem by degrading some habitats to the point where they cannot support the valuable species (McManus, 1997). Pacific Island countries now recognize that aquaculture provides one of the few long-term, sustainable, ways of deriving benefits from inshore fisheries resources (Williams, 1996). This view of aquaculture as a priority area for continued sustainable development was reinforced as part of a consensus member country statement arising from the 2nd SPC Fisheries Management Workshop in 1998:

> "*...Regional fisheries managers have focussed on establishing regimes to sustain inshore fisheries. This is supported by a strategy to divert demand and fishing pressure to alternative activities, mostly to offshore fishing and into aquaculture.*

Development of fisheries management varies from country to country reflecting the differing stages of economic development and levels of need. In some countries the need is to encourage economic activities and to generate income for rural villages; in other countries the need is to restrict or limit fishing. Yet in other countries the need to involve all stakeholders in the management system has evolved into community-based decision-making and control. There is now broad acceptance that marine resources cannot be managed in isolation from other users, or by one government agency so that an integrated and co-ordinated approach should be taken. In many circumstances, because of the smallness of the islands, an island system-management approach is the desirable option.

Aquaculture, as an alternative activity, is still at a preliminary stage of economic development in most PIC3, but is of enormous future significance. For aquaculture to realise its full potential to the economies of PIC in a sustainable way will require a considerable degree of international support. PIC have endorsed a strategy to harness and prioritise such support at the regional institutional level. Several PIC already devote significant national resources to this subsector and this trend will continue as benefits are realised..."

Advantages of Pacific Islands for Aquaculture: Pacific Island nations have many attributes that favour development of aquaculture and stock enhancement in the coastal zone. These include:

- o A great diversity of species associated with coral reefs that are in high demand for:
- o The aquaculture and seafood markets in Asia (e.g. napoleon wrasse, groupers, sea cucumbers, spiny lobsters, trochus, pearl oysters, giant clams, green snail);
- o The marine aquarium trade (e.g. clownfish, angelfish, hard corals, soft corals, giant clams); and
- o The pharmaceutical trade (e.g. algae, sponges, soft corals, sea horses).
- o Proximity to the major aquaculture and seafood markets of Asia-flight times are short enough to ensure that many species can be shipped alive to Asia.
- o Availability of suitable growout sites in pristine habitats-coral reef lagoons create the calm conditions essential for culture of many species. The favourable environmental conditions should be the opportunity to develop green label products to get better prices on the international market.
- o Geography that favours restocking and stock enhancement-most Pacific countries are small islands, or groups of small islands, surrounded by deep water. Cultured juveniles released into the inshore waters of island ecosystems cannot emigrate, and are therefore relatively easy to recapture.
- o A relatively inexpensive labour force-expectations for financial return on labour are low in many Pacific countries relative to developed countries.

- A tradition of working with marine resources-coastal communities are already familiar with the basic biology of many species.

Constraints to Aquaculture in Pacific Islands

Although the attributes listed above confer many advantages on the region for development of aquaculture and stock enhancement, there are also several constraints to such enterprises in the Pacific. Many of these have been identified previously by Uwate and Kunatuba (1983), Munro (1993) and Bell and Gervis (1999). They include:

- *Limited domestic markets.* Local markets for the fresh products of aquaculture in the Pacific are small, and with the exception of very limited opportunities in the restaurant trade, usually offer low prices. Thus any large-scale aquaculture development in the Pacific catering to the trade in seafood will depend heavily on export markets.
- *High added-value export markets targeted.* These are most often regional and fluctuating markets (e.g. live reef fish). Thus any aquaculture development will depend on the capacity to follow the market trends, and to fulfil the demand in time.
- *Transport problems.* The high cost of shipping in the Pacific adds considerably to the cost of producing and exporting aquaculture products. Poor internal transport services restrict opportunities to grow perishable products in remote locations, and limited international air connections inhibit continuity of supply to export markets. Transport arrangements dictate that species cultured for export need to be of high value and low weight. Alternatively, the products must be non-perishable, e.g. bêche-de-mer (processed sea cucumbers) or frozen, so that they can be shipped by sea.
- *Social and economic factors.* Many of the smaller island nations lack the infrastructure, capital and skilled labour required to implement aquaculture, particularly where hatcheries are involved. Sustained assistance from

developed countries is needed to implement and operate stock enhancement programmes until they become self-funding (Bell, 1999a). The traditional marine tenure systems in place in many countries (Ruddle et al., 1992) also add complexity to the process of negotiating access and tenure to sites for aquaculture.

- *Fragile habitats.* Coral reef ecosystems on many of the smaller island nations have evolved in a nutrient-poor environment. Additions of nutrients, e.g.through uneaten and undigested formulated diets for carnivorous fish in cage culture (Beveridge, 1987; Landesman, 1995; Stewart, 1997), can be expected to change the ecosystem in favour of algae and herbivores. Such changes are likely to be unacceptable, particularly to the tourist industry. This constraint is particularly relevant to lagoonal habitats, but would not apply to locations that have good flushing to the open ocean.
- *Freshwater is limited, except for the large islands of Melanesia, which have extensive river systems.* Prospects for freshwater aquaculture are thus limited. Even in areas with significant fluvial development, the indigenous freshwater ichthyofauna is generally unfavourable for economic culture and species for freshwater aquaculture have been imported.
- *Cyclones.* Countries in the cyclone belt can expect aquaculture installations to be damaged intermittently by large swells and strong winds.

Profitable aquaculture of penaeid shrimps and blacklip pearl oysters has now been established in some areas of the Pacific by commercial interests. Stand-alone enterprises producing penaeid shrimps for export markets are firmly established in New Caledonia and Fiji and were so in Solomon Islands until recently. These enterprises are applying technology developed originally in Japan, Taiwan Province of China and France, and now common place throughout the tropics.

A large, sustainable, industry for culturing pearls using the blacklip pearl oyster (Pinctada margaritifera) has been established in the Tuamotu Archipelago, French Polynesia, and on a couple of atolls in the Cook Islands (Fassler, 1995). In French Polynesia, the value of cultured pearls exceeds US$150 million per year. In Cook Islands, the industry is currently worth US$5 million and is the second largest source of revenue for the country after tourism. Black pearl farming in French Polynesia and Cook Islands, and shrimp aquaculture in New Caledonia represented more than 98 percent of the total value of aquaculture production estimated in 1996.

Blacklip Pearl Oysters: Small-scale culture of pearl oysters is under way in Fiji, Marshall Islands, the Federated States of Micronesia, Solomon Islands, Kiribati and Tonga. In some places, e.g. Kiribati, development is based on spat produced in hatcheries, whereas in others, e.g. Solomon Islands, development is geared towards finding ways that coastal villagers can catch and grow wild spat (Friedman et al., 1998). Current research is concentrating on assessing the economic viability of pearl farming in Solomon Islands, Fiji and Kiribati, and comparing growth, survival and pearl quality of oysters derived from wild and hatchery-reared spat.

Giant Clams: Small-scale enterprises in Solomon Islands, Palau, Marshall Islands, Cook Islands, Tonga and American Samoa supply five species of giant clams (Tridacna crocea, T. derasa, T. gigas, T. maxima and T. squamosa) to the marine aquarium trade (Foyle et al., 1997; Hart et al., 1998). Production of giant clams for enhancement of wild stocks is also under way in Solomon Islands, Fiji, Cook Islands and Western Samoa (Bell et al., 1997a; Bell, 1999b). Several of these countries also have the capacity to produce giant clams for the sashimi market in Okinawa, and as a live product for markets in China, Hong Kong Special Administrative Region and Taiwan Province of China (Bell et al., 1997b).

Sea Cucumbers: Research has commenced to assess the viability of producing sea cucumbers in hatcheries for enhancement of wild stocks. There are three steps in this process: developing methods for cost-effective mass production of juveniles, learning to release the cultured juveniles in ways that maximize their survival, and evaluating the economic impact of releasing cultured juveniles into existing fisheries.

Currently, the focus is on development of methods for the mass rearing of Holothuria scabra, H. fuscogilva and Actinopyga mauritiana, three of the most valuable sea cucumbers in the region. To date, ICLARM has demonstrated that H. scabra is relatively easy and cheap to rear (Battaglene and Bell, 1999), and that A. mauritiana grows relatively rapidly at high densities (Ramofafia et al., 1997). Initial research on H. scabra indicates that this species has much potential for stock enhancement.

Other Species: Technology for propagating and releasing cultured juveniles of green snail and trochus has been transferred to the Pacific through projects in Tonga (JICA) and Vanuatu (Austrailian Centre for International Agricultural Research, ACIAR). The Overseas Fishery Cooperation Foundation (OFCF) is also implementing a stock enhancement programme for both species in Solomon Islands. Production of the marine alga Kappaphycus is well established by coastal villagers in Kiribati and Fiji, and sponges are being cultured in the Federated States of Micronesia. Milkfish are being cultured as live bait for the tuna industry in Guam, and there is considerable interest in this activity by several other countries.

Policy and Institutional Framework: The independent Pacific Islands (Cook Islands, Federated States of Micronesia, Fiji, Kiribati, Marshall Islands, Nauru, Niue, Palau, Papua New Guinea, Samoa, Solomon Islands, Tonga, Tuvalu, Vanuatu) generally lack any specific provision for aquaculture in the legislation, but many include a statement about aquaculture in national development plans.

Whilst prospects for inland aquaculture are limited by geography, the custom of communal tenure of coastal marine areas may be incongruous with private-sector farm ownership in many areas, unless the development is carefully managed. Tonga is currently developing a comprehensive legislative basis for future aquaculture development, whilst the Cook Islands has adapted traditional systems into a legal basis for pearl farm management. Other countries are taking a more ad hoc approach and trying to adapt land-based systems of state leases for mariculture development, or encouraging development only by traditional reef custodians.

Fiji: Aquaculture was first started in 1953, when tilapia were introduced as a protein source for pig farming. The first directed efforts occurred after the United States Peace Corps and a JICA project assisted the Fiji government in developing freshwater aquaculture methods in the 1980s. Varied success was achieved with shrimp, Kappaphycus, oyster, mussel, Macrobrachium, carp and tilapia, but now major investment is promoted by the government through a Commodity Development Fund. Currently, there are three shrimp farms and hatchery; 20 milkfish ponds for longline bait; and one industrial, seven commercial and 215 subsistence farms and six hatcheries that produced 243 mt of tilapia in 1998. Eucheuma cottonii is cultured by a total of 182 farms, producing an estimated 1 500 mt in 1999. Experimental pearl farming is conducted, and there is one commercial farm that has been operating for two decades. There is also a giant clam hatchery, the Naduruloulou Government Freshwater Aquaculture Research Station, the Makogai Government Mariculture Research Station, and the University of the South Pacific (USP) Aquaculture Teaching-School Teaching Ponds (E. Ledua, 1999, unpublished data).

Guam: There was an increasing trend in annual aquaculture production between 1990 and 1995 (Clarke, 1997). In 1995, aquaculture production was 205 mt valued at US$1.6, derived mainly from the farming of tilapia, marine shrimps, Chinese catfish and milkfish. The potential for the development of other species (giant clams, top shell, striped mullet and groupers) is being explored by Guam-based research facilities.

Kiribati: Eighty hectares of mikfish ponds, originally set up to provide livebait for tuna pole-and-lining, have been operating for several decades. Kappaphycus alvarezii has been cultured in the Phoenix, Line and Gilbert groups for 15 years. There is an OFCF project on sea cucumber rearing and an ACIAR project on pearl farming (T. Tekinaiti, 1999, unpublished data).

Nauru: Milkfish been farmed for at least one century, but competition from introduced tilapia (O. mossambicus) occurred in Buada Lagoon and culture lapsed. It has recently been revived, with O. niloticus and milkfish being raised in 11 ponds (F. Alefaio, 1999, unpublished data).

Northern Marianas: Aquaculture is limited to raising penaeid shrimp and tilapia (Clarke, 1997). Annual production is estimated at 1 250 kg for marine shrimp, 200 kg for freshwater prawns and 4 200 kg for tilapia. There are fewer than three commercial farms. The total value of products was estimated as US$25 000 in 1996.

Nouvelle Caledonie: Shrimp production achieved 1 500 mt for the first time in 1998 (Labrosse et al., in press), after disease problems which occurred in the early 1990s. The value of marine shrimp production is about US$10 million. About 45 mt of oysters (Crassostrea gigas) are produced for the local market. Experimental culture of local oyster and giant clam is also in progress (Etaix-Bonnin, 1999).

Palau: The Micronesian Mariculture Demonstration Centre pioneered giant clam culture, as well as trochus and soft corals.

Papua New Guinea: Aquaculture started 40 years ago, with several aquaculture stations along the coast and highlands to encourage subsistence culture, mainly of Cyprinus carpio. There are 300 carp farms in operation. Trout were introduced in the 1940s, and the Kotuni Trout Farm was in operation from 1973-1984. There are three newer farms, but only two are currently operating, with a production of 15 mt. A hatchery was started in 1996.

Culture of barramundi (Lates calcarifer) has been started in Madang on the site of a failed 9 ha pond originally used for redclaw (Cherax quadricarinatus), but production has not yet been established (J. Wani, 1999, unpublished data).

Samoa: Farming trials were conducted on tilapia in 1954 by the SPC and on Macrobrachium in 1971 by the FAO, but were not successful. Seaweed, giant clam, green mussel and redclaw farming were also tried. A new national economic strategy promotes aquaculture, and this is being actively developed by the current AusAID village fisheries extension project using tilapia, mullet and giant clam (A. Mulipola, 1999, unpublished data).

Solomon Islands: There are two shrimp farms and several village-based enterprises rearing giant clams and hard corals for the aquarium trade and a demonstration black pearl farm in the Western Province. The ICLARM Coastal Aquaculture Centre has also developed methods for the propagation of sandfish (Holothuria scabra) and the capture and culture of wild postlarval coral reef fish for the aquarium trade. Between 1996 and 1999 there was an OFCF project on green snail & trochus (E. Oreihaka, 1999, unpublished data), but the future of the aquaculture-oriented Institute of Marine Resources of the University of the South Pacific is uncertain.

Prospects for Further Development

Some of the best opportunities for development of aquaculture in the Pacific are in the aquarium trade and live seafood markets (e.g. napoleon wrasse, groupers, sea cucumbers, spiny lobsters, trochus, pearl oysters, giant clams, green snail, abalone, crabs, clownfish, angelfish, hard corals, soft corals) and the pharmaceutical industry (e.g. algae, sponges, soft corals, sea horses) (Bell and Gervis, 1999; Bell, 1999c). In all cases, the products are of high value and can be grown in small areas with relatively simple technology.

Initiatives by FAO, the World Fish Center (ICLARM), the Center for Tropical and Subtropical Aquaculture (CTSA) and bilateral donors have concentrated on establishing the culture of pearl oysters outside eastern Polynesia, developing small-scale aquaculture enterprises for other species, and providing basic training in aquaculture and stock enhancement to fisheries staff in several of the countries.

The proceedings of the workshop entitled Present and Future of Aquaculture in the Pacific organized jointly by the South Pacific Aquaculture Development Project (SPADP) and the Japanese International Cooperation Agency (JICA) in Tonga in November, 1995 (Anon., 1996) presented the status of aquaculture in the region.

The continued expansion of aquaculture in the Pacific will depend on providing better methods of production for species currently under cultivation, and techniques for propagating and growing the "new" species described above (Bell, 1999c). These methods and techniques should be simple and flexible enough to be easily adapted to the context of the Pacific Islands environment and to the constraints of local and export markets. This approach should promote systems integrating fisheries and mariculture, with low cost methods of production). This should be associated with pilot commercial scale operations to test and demonstrate the economic viability of the methods proposed. This will need research coupled with assistance, training and education programmes.

In recognition of these needs, three organizations (SPC, ICLARM and USP) have recently joined forces to produce a "Regional Strategy for the Development of Aquaculture". Under this strategy, SPC will be the regional focal point for aquaculture and willconvene regular meetings of island nations to identify needs, determine priorities and put organisations and individuals in touch with each other. ICLARM will undertake long-term research to devise and test economically and environmentally sustainable methods for restocking, stock enhancement and farming; and USP will developdegree and diploma course components and aquaculture vocational training, and contribute to research through higher degree programmes. The other functions necessary for the expansion of sustainable aquaculture in the region, e.g. marketing, legislation, environmental protection and quarantine, will be integrated progressively through association with the Strategy.

This Pacific Islands Regional Aquaculture Strategy represents an opportunity to reinforce inter-regional cooperation based on research, training and information exchanges (including cooperation through NACA with Southeast Asian countries). It may promote more investment from Asia and better conditions for access to Asian markets.

THE FREE MARKET

Although marine fish, as residents of interconnected bodies of water that span the globe, are continually mobile and nearly impossible to localize or claim, as a resource for human consumption they have become a source of political debate and conflict between very localized states and groups of people. Currently, as the threatened state of global fisheries resources becomes increasingly prominent, the lines for disagreement, conflict, and negotiation are diverse and run the spectrum from international organizations to groups of different economic and social strata at the level of coastal communities. To people around the world of varying wealth, nationality, economic system, culture, and position of power, fish play a crucial role in their tastes, traditions, and needs for survival. This, more than what needs to be done ecologically to restore the world's fish resources, is what makes the situation so complex.

One effort to simultaneously resolve the seemingly disparate elements of the problem-the economic, the environmental, and the developmental-is the concept of sustainable development. In the context of fisheries, this concept entails being able to manage the pressures for the production and trade of fish, the strain this places on the resource base, and the needs of those people that depend on the resource for livelihood and food security to develop in an equitable manner. As ebbs or flows in one sphere impact the security of the others, solutions in no one sphere can be seen as representing a sustainable solution if the other two are neglected.

Of so much importance to so many people around the world, individuals, organizations, and states are searching for ways to alleviate the intense pressures on fisheries resources without completely sacrificing the interests of humans in using the resource for consumption and a source of livelihood. One current example of this is the proposal among states and international conservation and sustainable development organizations to use subsidy reform in the World Trade Organization (WTO) to pursue sustainable development objectives. While subsidies are a crucial contributor to the overfishing that occurs in many of the world's most important fisheries, the questions is whether or not such a reform can deliver on all three pillars of sustainable development. Subsidy elimination or reform in the fisheries sector through the WTO holds several legitimate possibilities for better resource use and alleviation of perverse economic incentives in fisheries products production that suggest advancement in important aspects of human development in developing coastal countries. This paper intends to explain, though, the limitations of this action and the possible threats to development and resource use in developing countries posed by such an endorsement of free market principles, especially given certain intranational realities of many of the most fish-wealthy developing countries.

Background on the Current State of Depletion in Fisheries Resources

The first step towards an accurate conception of current global fish-use, and possibly the least debatable aspect of the issue, is recognition of the precarious biological and environmental situation of the world's fish stocks in the face of increased global consumption of fish in the past 30 years. Since 1973, global consumption of fish has doubled. Growth in demand for fish as a source of protein in the developing world has accounted for ninety percent of this growth, while developed countries' share in world fish consumption has decreased. Fish is consumed most highly by far in Asia, and next highly in Europe. Fish consumption is generally higher in developed countries, but may play a larger role in food security in developing countries. At the same time, populations of fish in almost every part of the globe have declined to levels well below those they have historically and naturally held. The World Wide Fund for Nature (WWF) reports that 73 to 75 percent of the world's major fisheries are overexploited, fully exploited, or recovering from depletion. An article from the United Nations Food and Agriculture Organization (FAO) reports that, "Judging from known fish stocks and resources of traditional fisheries, the total marine catches from most of the main fishing areas in the Atlantic Ocean and some in the Pacific Ocean would appear to have reached their maximum potential some years ago." Global demand remains high, and the world's fisheries seem unable to support it in any sustainable way.

The link between increased global demand for fisheries products and the current level of stock exploitation is the dramatic increase in production that has occurred on an international level in the same time period. While 44 million tons of fish were captured for food in 1973, 65 were in 1997. Since the 1980s, developing countries have taken the lead in wild fish capture and now account for over 70 percent of global fish production. After such high levels in the last 30 years, though, wild fish production has stagnated; the state of exploitation in main fishing areas has actually remained more or less unchanged since the early 1990s. The world's fish stocks cannot continue to sustain such intensive levels of capture, a reality which has serious implications for both those on the production and consumption side of fishing.

The international trade of fish is also an important dynamic in the current situation. About 40 percent of world output by value of fish was traded across international borders in 1998. Since the early 1970s, the direction of fish trade has experienced an "about-face:" developed countries were net exporters of 818,000 tons of food fish in 1973 but, in 1997, were net importers of over four million tons of food fish. More than half of fishery exports originate in developing countries, while developed countries account for 84 percent of world fishery product imports. Fish products are vital exports for many diverse economies, but the top exporters of fish and fish products are Thailand, the United States, Norway, China, Denmark, and Taiwan (in that order). Imports into the three largest importing blocs, Japan, the EU, and the United States, accounted for over three-fourths of the world market in international fish trade in 1994. As will be addressed, although fish trade is important for many, it is hardly on equal levels or terms, and benefits are rarely bestowed on those to whom the resource matters most for food and livelihood security.

Fisheries: An Open Access Resource

Due in large part to the way it has traditionally been managed, marine fish has usually been seen and used as an open access resource, an in many ways endemic characteristic which has been the base of its dynamics of use, management, and exploitation. The term refers to a situation where no single user has to pay for access to a resource, nor does that user have exclusive rights to the resource or the right to prevent others from using it. Physical characteristics of the oceans, and the fish they contain, such as scale, interconnectedness, and mobility of fish stocks, contribute to this open access nature. As such, they have usually defied national borders and been logistically difficult or impossible to monitor and govern. Theoretically, this can lead to a "tragedy of the commons," where "the open-access nature of the fisheries allows more and more people to enter the fishing sector, which leads to the economic (and possibly biological) over-exploitation of the resources." Even after the resource begins to decline, producers will continue to increase their efforts until operating costs can no longer be covered. By this time, the stock of the resource may be seriously depleted. The common-access nature, which has long been a characteristic of fisheries, is the foundation on which many of the causes of current problems in the sector are based.

Root Global Causes of Fisheries Resource Depletion

The simultaneous, and highly linked, realities that too many fish are being caught and the global demand for fish is unsustainably high are largely seen to be the result of four root causes: population growth, growth in incomes, urbanization, and advancements in fishing technology.

Population growth has contributed to an increase in demand for fish by increasing the number of people who demand fish. Income growth has played a role because, as incomes increase, wealthier people's tastes change from preferring low-value to high-value fish; more and more low-value fish are then diverted "from poor people's plates into the stomachs of expensive carnivorous fish, which will in turn end up on the plates of wealthier people." A 2003 report by the World Fish Center found that population growth and income growth were "widespread structural phenomenon," especially in the developing world, which drove increased demand and consumption in the last three decades.

Urbanization has also impacted the structure of fish demand because, as the "rapidly expanding" group of poor people living in larger cities throughout the world increases, low-value fish is their only affordable source of animal protein. On the production side, developments in technology that have allowed fishermen to catch more per unit input and seek out fish in deeper waters have kept production costs down and allowed production levels to increase. Roman Grynberg and Martin Tsamenyi, for instance, cited constantly improving technology as a fundamental cause of fish stock depletion. The interaction of these structural forces of increasing incomes, populations, and urbanization, as well as advancements in technology, have lead to the over-use of the resource, which is, as is currently being witnessed, limited in its capacity to respond to and accommodate them.

RESPONSES TO THE CHALLENGES OF FISHERIES USE: MANAGEMENT AND SUBSIDIES

The very reality that the root causes of the over-use of fish stocks are structural makes it nearly impossible to completely alleviate the pressures of rising demand, increasing abilities to pay for high-value species, and more effective ways to capture fish on an open-access resource. But people, organizations, and governments have responded to these realities in many ways, their responses sometimes improving the situation, but often exacerbating it. Two of the most prominent and important responses have been subsidies and management schemes. The first, subsidies, occur when trade measures directly influence the income or reduce the costs of production for industry. Management schemes, the second, encompass the ways in which fisheries are governed and monitored. In practice, the way this is done often highlights a lack of effective management. Subsidies and management, which can be seen as responses by individuals, groups, or governments to the root causes of fisheries depletion, are also the areas where the most opportunities for change in the way fisheries resources are used exist.

Effective management regimes are considered by many to be key in the sustainable use of fisheries resources. Theoretically, effective management encourages a sustainable use of resources through providing conservation regulations and incentives for responsible use of the resource. They are especially beneficial under the pressures of international integration; responses by producers to the economic incentives from trade liberalization, for example, can be better regulated when effective management schemes exist so that over-exploitation and destruction are less likely to occur. Unfortunately, the prevalence of such management has been scant in real-world fisheries. As the United Nations Environment Programme (UNEP) reports, "very few fisheries management systems have demonstrated the ability to keep catches below levels that put pressure on stocks." As noted in a study by the World Conservation Union (IUCN), "national and international progress on trade expansion and liberalization outpaces progress on fisheries management and articulation of sustainable development strategies." Fisheries management, crucial to sustainable use of the world's fisheries, has been difficult to implement effectively and has often left fisheries open to influences from other sources such as state trade measures.

Subsidies, one of these trade measures, have played a proven role in aggravating the causes of fisheries depletion. Theoretically, subsidies can lead to overuse of a resource by altering the behaviour of producers. Revenue-enhancing or cost-reducing subsidies, such as those that will be examined in this paper, increase marginal profits at each level of effort and therefore theoretically lead to an increased overall effort. This may have the short-term effect of creating additional economic rent for producers, but either new entrants or increased effort by existing producers, also encouraged by the subsidy, will shift the level of effort so that rent is eventually dissipated. Assuming that the management program does not effectively impose a sustainable level of resource use, cost-reducing and revenue-enhancing subsidies will push the level of overcapacity and overall effort even further than would an open-access resource in the absence of subsidies. The main theoretical effects of cost-reducing, rent-increasing subsidies are increased capacity, a delay in exit by producers from the sector, and increased effort by producers. These effects can cause the overuse of a resource, which can have many other significant impacts such as threats to food, livelihood, and resource security.

FISHING SUBSIDIES

Types Used by National Governments

In the real world of governments, fishers, fisheries resources, and the incentives that link them all together, a subsidy can take many forms. Fishing subsidies currently account for large portions of many industrialized country budgets; a 1998 World Bank study estimated annual budgeted expenditures on global fisheries subsidies to be between $14 and $20 billion, and that subsidies account for 20 to 25 percent of the sector's revenues. One type of subsidy, domestic assistance programs, form a clear example of government endorsement of the fishing sector. These programs are defined as budgeted assistance to fishing production, and often enhance operations and capacity. Japan, one of the biggest subsidizers of fishing activities in the world, has used domestic assistance programs to recruit young fisherman, give aid to fish cooperatives and boat owners, support marketing, and encourage price stabilization.

A 1998 World Bank report estimated that Japan budgets $270 million a year for these types of programs, and a total of $750 when aid for fishing vessel insurance is included.

The European Union (EU), another major subsidizer, supports such structural programs as fleet renewal and modernization, processing and marketing, aquaculture, maintenance of port facilities, and genetic product promotion in their fisheries sector. In addition, it pays for market supports to the sector in the form of a minimum import price program and measures to support price floors. Costs of these programs to the EU in 1996 were estimated to be about $530 million.

While estimates of exactly what types of domestic subsidy programs the Chinese government uses are hard to determine due to scant data, they have clearly aggressively promoted the expansion of their fishing sector in the last 15 years (and have become the world's leading producer) and are estimated to budget between $500 and $750 million annually for domestic subsidies. While other countries subsidize their domestic fishing industries, they have comparatively small budgets for such programs as so do so on a smaller scale. The shear magnitude of the domestic subsidies in industrialized countries suggests the possible impacts on the behaviour of fishers their presence may cause. A less obvious, but yet potentially just as influential, type of subsidy occurs when a country pays the access fees for their fleets to fish in foreign waters. Unlike domestic subsidies, these payments are not made directly to fishers or even domestic programs, but to the foreign governments who have jurisdiction over the fish stocks. The EU is one of the major subsidizers of distant-water fishing through its numerous access agreements with developing countries, mostly in West Africa. For example, the EU signed a five-year access agreement with Mauritania in 1996 that lifted an EU embargo on fishery imports from Mauritania and specified EU payments of almost $350 million over five years to the Mauritanian government. In return, the Mauritanian government granted increased access for EU fishing vessels in their waters, authorization of higher EU total harvests, and the specific allowance of EU-directed fishing of highly valued squid and octopus. The monetary benefits to the "cash-strapped" nation were described as a "windfall;" in consequence, the number of eligible EU boats rose from 165 to 240 and allowable EU harvests from 76,050 to 183,392 tons. The EU has similar agreements with other West African nations, as well as others in East Africa, the Indian Ocean, and South America. Another distant-water fleet subsidizer, Japan, currently budgets about $200 million a year on distant water arrangements with foreign countries and foreign fisheries assistance programs (which often help secure fishing rights). These payments went mostly to developing countries in the Pacific Ocean and elsewhere. The subsidization of distant water fleets from industrialized nations can have serious effects on the coastal communities of the developing countries with whom these agreements are typically made.

Possible Impacts of Perverse Subsidization

The impacts of government subsidies of fishing are at least as complex and numerous as the types of subsidies themselves, but in the context of sustainable development those related to market access due to trade measures and resource, livelihood, and food security are the most relevant. The trade implications of the subsidization of the fishing industry arise when the lower costs producers see as a result of the subsidies repress world prices for fish products. Those producers, who are often from developing countries that cannot afford subsidies, are then excluded from the market because they cannot afford to produce at the world price.

For example, when Japan subsidizes its fleets to fish in the waters of Chile, Japanese fishers bring the fish back to Japan and sell them on their domestic market. Chilean fishers try to enter the Japanese market and sell these same Chilean fish to Japanese consumers, but are kept out of the market because their costs are higher and the prices on the Japanese market are too low. In effect, Japanese fishers push Chilean producers out of the market for the sale of their own fish. This example illustrates the way in which subsidies in the fishing sector can exclude those who are outside the national subsidy regime from participating in the market.

Due to numerous examples where the presence of subsidies seemed to increase the capacity in the sector and, in turn, deplete available resources, fishing subsidies have also been blamed for a tendency to encourage overcapacity and overfishing. The case of stock depletion in Canada's Northwest Atlantic fisheries in the 1950s and 1960s illustrates how subsidies can increase capacity and lead to depletion. Between 1954 and 1968, Canadian subsidies increased the capacity of the Northwest Atlantic offshore fishing fleet by more than 18 times, creating twice as much capacity as could capture fish if resources were to be used sustainably. In 1970, the Canadian government "acknowledged explicitly that its subsidies for vessel construction and modernization over the previous two decades had led to the rapid expansion of larger vessels, creating serious overcapitalisation." An Organization for Economic Cooperation and Development (OCED) study reported that subsidies also contributed to overcapacity in the fishing sector in many member states: New Zealand in the 1970s and early 1980s, Spain's Galician fisheries in the 1980s, Norway in the 1960s, the EU during the 1980s, and the United States (US) in the late 1970s and 1980s. These examples of the history of fishing fleet subsidization in industrialized countries and the resulting depletion of their stocks illustrates how overcapacity in a sector can encourage overuse of a fragile resource and lead to its exhaustion, posing threats to the surrounding ecosystem and human environment.

In addition to threatening the resources on which humans depend for their incomes, fishing subsidies may pose direct threats to crucial aspects of human and national development such as livelihood and food security. These threats are seen most strikingly in the interaction between developing nations and the industrialized countries with whom they sign access agreements. Frequent realities of EU-West African access agreements, for example, include a disregard of African fishing regulations and international agreements by distant water fleets, an inability by African states to monitor activities or catch levels of EU fleets, a lack of direct linkage between access payments and money spent on fisheries-related activities, a lack of enforcement of management efforts by African governments due to a fear of losing compensations, and the challenge of depletion of valuable fisheries stocks. There are also often many adverse effects on African coastal communities such as threats to the livelihoods of artisanal fishers and food security of those who depend of local fish stocks for protein intake. While EU payments often constitute important sources of revenue for West African governments, the security of fishing resources, vital to so many aspects of these countries, are often threatened.

The case of one West African country that signed an access agreement with the EU, Senegal, illustrates how subsidies and trade agreements can lead to threats on food and livelihood security. The Senegalese government initially implemented subsidies as a way to encourage fish production to supply domestic consumers with a protein source; fish accounts for 75 percent of the population's protein needs. But the use of subsidies has since become incompatible with this goal since the end result has been a re-orientation towards encouraging the exportation of fisheries resources and agreements with the EU for access to Senegalese fisheries. The combination of domestic export subsidies, which lower the costs of production for those fishers who chose to export, and trade advantages granted under the Lomé Agreement of 1982, which authorized customs exemptions to products originating in African, Caribbean, and Pacific (ACP) countries and increased the competitiveness of Senegalese piscatorial products in the European market, have encouraged small-scale fishers in the country to both intensify their efforts in response to international demand as well as orient themselves towards the export market.

These developments in subsidy policy, along with several fisheries agreements that have been signed with Japan and the EU, have threatened livelihood and food security in Senegal. The impact of growth in subsidy-driven exports has lead to a "breakdown in the supply of cheap protein to the population," since local producers began to switch their activities toward the capture of species of high market value in European economies rather than those affordable to local consumers. The increasing scarcity of fisheries resources has also affected the traditional, artisanal fishers, whose techniques, which require low technology and employ many people, are beneficial for local resource and employment security.

Since fish have become scarcer, foreign fishing fleets have become a larger presence, and local fishers lack the equipment and fuel to travel farther out to sea where the remaining catches are located, they have been forced to make deals with European or Asian boats to remain employed. In return, locals allow the foreign boats access to areas originally reserved for their needs and use. This trend threatens local control of the resource as well as livelihood security. In a country where fish provides a vital source of sustenance for the local population, in terms of both food and employment, the way in which subsidies encourage a reorientation of resources to foreign markets is a matter of much concern for the sustainable development of the country.

A ROLE FOR THE WTO IN ACHIEVING SUSTAINABLE DEVELOPMENT OBJECTIVES?

The Environmentalist-Free-Trader Alliance

As the potential threats perverse subsidies pose to the world's fisheries and those who depend on them become increasingly evident, many in the international community are seeking ways to link reforms in subsidies to those clearly also needed in management and resource use. One option that has recently become popular among environmental and sustainable development organizations and several fish-producing nations is the pursuit of subsidy reform through the WTO. This endorsement is significant in several ways. First, the connection between the liberalization of trade through the WTO has not traditionally been a cause of environmental organizations. As Grynberg and Tsamenyi note, "What is peculiar is that environmentalists, who for so long focused on economic and population growth as the cause of resource depletion, should now enter into alliance with the free traders and focus on a symptom rather than what they have traditionally seen as the root cause of the problem."

Environmentalists have also conventionally highlighted the numerous potential conflicts between the liberalization of international trade and the use of the world's resources. In the context of fishery reform, however, international conservation organizations such as the WWF and the IUCN have expressed that their organizations hold several aims parallel to those of the WTO. A related second shift this "alliance" illustrates is the progress of thought concerning the role of the WTO and trade liberalization. Instead of seeing measures such as subsidy reduction purely as ways in which to influence economic factors such as market access and prices, these same measures are now being seen as capable of influencing areas of sustainable development such as the environment and human and national development.

Since its creation in 1995, the WTO has included sustainable development as one of its objectives along with the liberalization of international trade, but the linkages between the two objectives have not always been clear. The WTO promotes a world trading system based on the principle of comparative advantage, suggesting that, "In a trading system in which all nations remove protectionist measures and other barriers to market access, the net benefits to all should increase." According to the WTO, subsidies are trade-distorting measures that impede the operation of the principle of comparative advantage. What the WTO, a forum for negotiation of trade measures, cannot assure is to whom the economic rents from this process go or how resources are used in the process of trade and production.

Background on the Place of Fisheries in the WTO

Fisheries, their trade, and the subsidization of their production have yet to find their places in the WTO. Due to conflicting interests during the Uruguay Round negotiations, the issue of fisheries subsidies was not resolved at the Organization's inception, and fisheries subsidies were included under the Agreement on Subsidies and Countervailing Measures (SCM). This agreement covers all goods except agriculture and attempts to place restrictions on the power of WTO members to provide subsidies to their industries. As UNEP reports, since 1995 a "significant change in the political context of the issue has taken place.

The impetus for inclusion of fisheries subsidies on the international trade agenda has increased substantially, primarily because it has been treated not only as an issue of efficiency and equity in international trade but also as an issue of protecting natural resources from depletion." Momentum for getting fishing subsidies put on the trade agenda increased in the late 1990s, in particular at a special session of the General Council of the WTO where Iceland, with the support of nearly 20 other countries, proposed that members agree to eliminate those subsidies that contribute to overcapacity. The desire among many WTO members to eliminate capacity-inducing fishing subsidies set the stage for an increased role for fisheries in future WTO negotiations.

In anticipation of the fifth Ministerial Conference in Cancun, Mexico of September 2003, a group of states calling themselves "The Friends of the Fish" lead the attack on fishing subsidies and their apparent threats to world trade and sustainable development. The group, which formed prior to the Seattle Ministerial Conference of 1999, successfully campaigned for inclusion of a paragraph in the Conference's draft declaration but failed to see an agreement be reached as the meeting broke up without a final ministerial declaration. Australia, Chile, Ecuador, Iceland, New Zealand, Peru, the Philippines, and the US are the leaders of the "Friends." Japan and South Korea, which have large subsidy programs for their fishing fleets, have argued that fishing subsidies should not be treated in terms of having trade-distorting effects and question the linkages others claims the subsidies have with resource exploitation. The EU is between the two sides; traditionally having sided with Japan and South Korea, it has recently been less defensive of its subsidies. Hoping for better results in the Cancun negotiations of 2003, the "Friends" then submitted a proposal to the WTO in April of 2002 outlining the "potentially harmful trade, developmental and environmental effects of subsidies to the fisheries sector" and suggesting the need to address these effects. While negotiations at Cancun also failed to deliver a consensus on the issue, the group and their allies continue to push for reforms through the WTO that would eliminate what they consider perverse subsidies in the sector.

Input of International Organizations

The call for the reform of fishing subsidies has come not only from member states of the WTO, but also from international non-governmental organizations. Economic organizations that have become involved include the OECD, which undertook the first study to measure fisheries subsidies in OECD countries in the early 1990s, and the Asia-Pacific Economic Cooperation Forum (APEC), which collected data on fisheries sector support programs in member governments. The FAO has played a crucial role in putting fishing subsidies on the radar; by their account they "sounded the alarm about the negative effects that subsidies were having on capture fisheries" in 1992, which "led governments of major fishing nations and various Intergovernmental Organizations to look into the issue." These organizations stress the link between fishing subsidies and their ability to cause perverse economic incentives and overcapacity.

Organizations with conservation and sustainable development agendas have also added their voices to the criticisms of fishing subsidies, and many have iterated a possible role for the WTO in the attack.

The WWF, for example, has published much literature emphasizes the link between government subsidy payments and overfishing and, in one essay, suggests that "The World Trade Organization (WTO) today has an unprecedented opportunity to help improve the environmental and economic health of the world's oceans by disciplining harmful fishing subsidies."

The IUCN has also described a role for subsidy reform and the WTO in the achievement of fishery use within sustainable development aims. One publication examines possible "synergies" between "international trade, sustainable fisheries, the environment, and sustainable development." By focusing on the causal factor of subsidies in overfishing and resource depletion, these organizations have created a bridge between a trade measure and the state of natural resources, thereby creating a role for an economic organization such as the WTO in the pursuit of sustainable development aims.

The Sustainable Development Concept

The context into which governments and organizations are currently trying to put subsidy reform, that of sustainable development, pulls together concerns that are a combination of economic, developmental, social, and environmental. The concept of sustainable development became prominent at the 1992 United Nations Conference on Environment and Development, which defined the idea as "meet[ing] the needs of the present without compromising the ability of future generations to meet their own needs." With its incorporation of a commitment to poverty eradication, reductions in disparities of standards of living, and environmental protection as an integral part of the development process, sustainable development "challenges the system to reconcile trade objectives with broader environmental, social, and cultural imperatives, and also raises concerns as to the issue of what constituted effective and legitimate development, rather than measuring development in narrow terms of economic growth."

Sustainable development in terms of fisheries includes meeting the economic, social, and cultural needs of coastal communities, addressing the concerns of developing countries to maintain revenues from trade that are necessary for development, monitoring the health of fish stocks, and surveying the surrounding situation of biodiversity. This perspective encompasses much more than either the strictly economic side-benefits to increased fish production and trade, or the conservation side-sizes of fish populations.

Reform of fishing subsidies, an issue that seems to have become a nexus of economic, environmental, developmental, and social interests, is defined by many as a potential 'win-win-win' situation for all concerned.

According to the WFF, "the WTO and its members have been promising for years to negotiate new rules that deliver true 'win-win-win' outcomes for trade, the environment, and sustainable development," and the fisheries subsidy issue presents the "first real" opportunity to move toward this objective.

In suggesting that such a situation exits, 'win-win-win' outcome proponents conjecture that the same measure, here the elimination of perverse fishing subsidies, could deliver favourable benefits to trade by increasing market access for producers, the environment by reducing the capacity of global fishing fleets, and sustainable development by assuring resource and livelihood security to those in developing countries.

The Argument for the Use of Subsidy Reform in the WTO to Achieve Sustainable Development Objectives

The argument for proposed changes brought through the WTO, that many view as having the potential to deliver this 'win-win-win' outcome, are based on three main justifications of subsidy elimination: trade, sustainable development, and the environment. The trade foundation of the argument deals with the way in which cost-decreasing, rent-increasing subsidies may deflate prices on the international trading market because of lower costs for producers due to government support that absorbs a portion of costs.

The submission made by the "Friends of the Fish" to the WTO stated that subsidies in the sector caused distortions to trade through their "effects on prices and production levels," and that the transfers "significantly distort trade in fisheries products by lowering costs of subsidized production, increasing the volume of subsidized production and thus distorting competitive relationships in markets for fisheries products."

Concerns for the impact of such distortions are also voiced by organizations, mainly in the context of how they might adversely affect the economies of developing countries. An IUCN report suggests that, for many developing nations, "trade liberalization offers possibilities for addressing declining terms of trade," and that efforts to reduce subsidies could have the positive impact of enhancing " export opportunities for countries currently shut out of export markets due to subsidized competition." These factors are important for developing countries in particular because of questions of equity in trade; as Charlie Arden Clark of the UNEP expressed, those best able to subsidize themselves (wealthy developed countries) will best be able to obtain the resource and trade it on the world market. The importance of fishery product exports to the economies of many developing countries is also an important part of the argument; the FAO highlights how "exports of fishery products are vital to the Least Developed Food Deficit Countries, in terms of export revenue, employment and income generation." The potential benefits to elimination of perverse subsidies in the fisheries sector such as increased market access for non-subsidized competitors comprise part of the argument in how subsidy reform might form an important part of sustainable development. The potential positive impacts of addressing fishing subsidies in the context of an international trading system could extend beyond those that are purely economic, though, and contribute to several important aspects of human and national development. Due to the way in which they can threaten food and livelihood security, the effects of access agreements and the foreign fleets they bring into the waters of developing countries are being targeted in particular as a way to move toward development aims. An IUCN report notes that one positive impact might be to "Provide greater opportunities for domestic fisheries industries in developing countries by reducing subsidies to foreign distant water fleets."

If access agreements were eliminated, there may be less competitive pressure frcm foreign industrial fleets, which "effectively compete for fish with local artisanal and industrial fisherman, thereby threatening their livelihood security, and undermine the economic development of poorer West African nations." The presence of foreign fleets and the governmental revenue they bring are also of little development assistance to local fishing communities. As Aimee Gonzales of the WWF explained, the fees that developing country governments receive as access payments are not used to access capacity or monitor stocks, measures that might enhance local capacity to develop a sustainable fishing sector. She also described how there is little "trickle-down" effect from the presence of the industrial fleets in the coastal communities; since the big, foreign boats have the ability to process their catches on-board rather than use local processing plants, they contribute little to the local economy. The elimination of access agreements may, in some cases, also help increase food security by increasing the share of fish that go to foreign fleets for export. Developing countries hypothetically stand to gain much from a decreased presence of foreign fleets in their waters.These potential gains for development in eliminating foreign access to fisheries of developing countries have inspired proposals from several organizations for reform. An IUCN report, for example, challenges the agreements to be considered subsidies under the WTO SCM Agreement, as well as proposes a new international agreement that "includes provisions prohibiting the payment by governments of any part of the costs of access to fishery resources." The WWF also recommends that the WTO expand its definition of "subsidy" to include the access agreements, and to address the issue of sustainability by "considering whether subsidies contribute to excess capacity, over-fishing, or unsustainable fishing practices." Under such propositions, a subsidy may be prohibited under WTO rules, even if it does not cause trade-distortions, as long as it is seen as being counter-active to sustainable development goals.

The environmental impacts of cost-reducing, rent-increasing subsidies are looked at mostly in terms of how the subsidies contribute to fishing fleet capacity. The link between subsidies and overcapacity is crucial in understanding both the role subsidies have played in causing the current state of resource depletion and what role they might play in the future in the pursuit of more sustainable fishing behaviour. According to an UNEP study on the effects of subsidies in the fishing sector, "The key issue for analysis and assessment in regard to the linkage between subsidies and overfishing is whether and to what extent fisheries subsidies contribute to fleet overcapacity." Fishing fleet overcapacity has been identified as a major threat to fishery resources, and global fishing fleets are currently estimated to be up to 250 percent greater than needed to catch what the world's oceans can sustainably produce. Carl Gustaf Lundin, Head of the Marine Programme at the IUCN, emphasized the need to reduce capacity in order to achieve fisheries resource security. Because of the threats they pose to the environment, governments and organizations suggest that capacity-inducing subsidies be prohibited by the WTO. An IUCN report proposes that various modifications be made to the WTO SCM Agreement to "ensure that all capacity inducing subsidies are covered by the agreement." The WWF and the "Friends of the Fish" also suggest that the WTO rules address the issue of subsidies that contribute to excess capacity and contribute to fish stock depletion. These groups hope that, if subsidies that increase fishing fleet capacity are targeted and eliminated, fish stocks may be able to recover in the face of decreased fishing intensity.

While most of the argument for the link between the elimination of perverse subsidies and certain sustainable development goals has been based on these three objectives, some feel that a role for the WTO may in itself add to progress. They believe that the WTO can potentially increase political action in the area of fisheries and improve other important aspects that need reform, such as fisheries management. Aimee Gonzales at the WWF explained that, since subsidies are where the politics are, action through the WTO might spur "parallel" action among other bodies to create good management regimes and, thus, sustainability. Hugo Cameron of the International Center for Trade and Sustainable Development (ICTSD) agreed with this idea, noting that the fact that the subsidies issue is being brought up at the WTO will give fisheries "air time." He also felt that recent improvements in the EU's Common Fisheries Policy were due in part to pressure from governments and environmental groups for reform inline with WTO rules. The "Friends of the Fish" submission to the WTO describes a role for the WTO in the context of a situation where action is clearly needed: While many factors contribute to depletion of the world's fish stocks, harmful subsidies are one important factor within the WTO's competence to address....As the forum in which global trade rules are negotiated and enforced, the WTO has a unique and important role to play as part of a multi-layered global response. The WTO's mandate, experience, expertise, and pre-existing framework of rules and dispute settlement mechanisms are critical to dealing effectively with the trade-related aspects of this problem. An integral justification of the proposed role for subsidy reform through the WTO has become the benefits the WTO can uniquely contribute to the situation due to its international political importance and prominence.

Evaluation: The Implications of WTO Subsidy Reform as a Method for Achieving Sustainable Development

Given the complex and far-reaching impacts organizations and states conjecture for fishing subsidy elimination through the WTO, it will be difficult for actual reforms to live up to their model. Certain benefits seem assured. On a net basis, trade liberalization can increase the economic gains of a developing country by increasing their market access. And, in a vacuum, the elimination of subsidies can decrease productive capacity and leave more resources available for the use of those who were formerly excluded. Sustainable development, though, is a much taller order. It means assuring that certain aims can be met in the context of forces such as trade and resource use. Whether or not WTO action on fishing subsidies can achieve these aims is the question. The ability of subsidy elimination to decrease overfishing by subsidized fleets or create market access where there perhaps was none before is not under debate-what are are the implications and possible benefits of this action.

Regarding the environmental sphere of the argument-the use of subsidy reform for conservation aims-the security of fisheries resources has many factors beyond that of the subsidization of fleets. While the theoretical reasons that support the role for cost-reducing, rent-increasing subsidies as a potential cause of increases in capacity are strong, and many empirical cases have illustrated instances when increases in such subsidies played a pivotal role in overcapacity, the extent to which they have done this is still under debate. The importance of management in the relationship between subsidization and overfishing has been found to be incredibly significant since effective management may alter the impact of subsidies on resource use. If fisheries are well managed and impose regulations and/or quotas, additions to capacity may not result in additional effort.

The OECD Committee for Fisheries reports that the effects of subsidies on fisheries resources are theoretically different depending on the type of management regime present. As could be expected, the effects of subsidies on open-access-type management regimes include overexploitation and increased capital and labour being attracted to the industry. In contrast, under an effective management scheme, the same subsidies had no effect on resource sustainability. By conjecture, if effective management schemes were also present in developing countries, the degradation of their resources in the face of highly subsidized fleets may not seem so imminent. Theoretically perverse subsidies clearly create overcapacity and resource degradation in many cases, but do not stand alone as the causal factor of current fisheries depletion.

The validation for subsidy elimination based on trade illustrates the complexity of achieving a true 'win-win-win' situation, especially when the interests of one of the main supporters of this view, the "Friends of the Fish," are examined. The group, described by Hugo Cameron at the ICTSD as a group of relatively competitive fishing countries, is lead by the US and New Zealand, fishing nations that subsidize their fleets relatively little. Support by these two countries in particular seems to be rooted in their economic concerns over the competitiveness of their fishing fleets. The US government apparently blames the lack of profitability of its fleets on the low world fish market prices, which it sees as being held down by the subsidies in distant water fishing nations. Eliminating subsidies in the EU and Japan, for example, would theoretically increase prices and improve the US's market share. According to Grynberg and Tsamenti, environmentalists who support trade liberalization as a means to sustainable development and environmental security are in effect supporting US trade interests. To the "Friends of the Fish," eliminating subsidies in the fishing sector is a way to increase the competitiveness and productivity of their fleets.

What the motives of the US and the "Friends" highlight is that there is a significant endorsement of free market principles in the attack on fishing subsidies, principles that, if implemented, might unleash threats on areas crucial to the pursuit of sustainable development and resource security. An IUCN report suggests the possibility that "the free functioning of comparative advantage in trade relations may not per se encourage the best long-term allocation of resources in the fisheries sector." This possibility becomes especially acute when the lack of effective management of developing country fisheries is recognized. While, in many industrialized countries stocks are monitored and assessed and few fisheries remain purely "open access," developing countries lack the capacity and funds to effectively manage their fisheries. The WFF reports, for instance, that "Traditional management methods like surveillance patrols, ecological monitoring, or measures to address pollution from both land and sea have met with only limited success because they are simply too costly for the region's cash-strapped institutions." This asymmetry in management between the North and the South may actually encourage fishing in developing country waters, where management is more lax. This idea is supported by economic theory, which warns of a comparative advantage that can seem to be present for the production of common property resources in the South due to their less well-defined management or property rights schemes.

As described by the economist Graciela Chichilnisky, "lack of property rights alone can create trade" between two countries that are otherwise identical since the ill-defined property rights in one country alone will cause the overuse of a resource in that country. Fisheries biologist David Boyer explained that increased market access for developing countries may indeed increase pressure on fisheries as fishers in these countries scale-up their fleets in response to international demand for high-valued fish.

In his opinion, market liberalization may well "shift pressure to greater fishing for trade from countries that have little environmental protection and the inability to enforce international treaties or their own national laws." He also notes that, without "deliberate domestic policies to facilitate restructuring" in the face of trade liberalization, "liberalization may, at least in the short and medium term, actually work against growth, employment, poverty alleviation, and other components of sustainable development." Liberalization of fish trade, one of the impacts of subsidy reform, may only continue the current pattern of resource overuse that threatens the development and resource security of developing coastal nations, or, at worst, increase this overuse.

Better management of fisheries in developing countries increases the chances of "winning" on several development fronts since it provides some assurance of how the resource would be used after the elimination of subsidies in addition to "less," or "more in line with market principles," as the subsidy elimination argument suggests. As described, free trade of fisheries resources may put pressure on the resource and do little to assure livelihood security for local communities. Effective management schemes could enforce quotas and regulations, assuring livelihood security for future generations. Current livelihood security in many developing fishing nations further depends on the lessening of competitive pressure from distant-water fishing fleets from developed nations.

Subsidy elimination would very likely lessen the number of EU or Japanese boats in these nations' waters, but this change says little about which boats, such as those of Iceland or Argentina, known to be very efficient, might enter next. Even when subsidies are removed, high global demand for fish will still exist, as will the reality that a large percentage of the world's fish are located in the waters of Western Africa and Western South America. In my opinion, management is one of the only likely mechanisms that can relieve such market pressures on the resource. Food security may be improved little by removing subsidies if, as the trade scenario described suggests, producers still have large incentives to focus their efforts on high-value fish for export to developed countries. If management were effective, though, catches by industrial or foreign fleets could be limited and some resources could be reserved for local population use, a method that has been proposed in Mauritania. Management of fisheries in developing countries is not only a key aspect of resource sustainability, but also a necessary guard against potential threats of subsidy elimination. Organizations and the "Friends of the Fish" argue that certain characteristics of the WTO, such as its prominence in political and economic arenas, may help sustainable development aims by leading to a chain reaction of reforms in other areas. This argument's merits aside, the international scale of the WTO, whose rules "are intended to help ensure that trade flows as smoothly, predictably, and freely as possible" at an international scale, poses potential threats to interactions at other levels. For example, as the push for action on fisheries issues at the WTO increases, those engaged in fisheries agreements at the regional level are raising concerns about potential conflicts between regional agreements and international WTO rulings. Regional Fisheries Organizations (RFOs) adopt their own trade measures between states in a region in order to achieve purposes such as compliance with conservation and management regimes.

RFOs can ensure cooperation through requirements for documentation, prohibiting landings from non-compliance vessels, or implementing trade measures. Given their aims for effective management, RFOs present an opportunity for reform in an arena crucial to fisheries resource conservation and largely left out of the impacts of subsidy reform. But, as Richard Tarasofsky forewarns, RFOs, which "seek to discriminate in favour of goods produced in accordance with their conservation and management aims, are potentially in conflict with these [WTO] provisions," which mandate equal trade treatment of like products. A winning situation at the WTO may be a loss at the regional level if WTO rules that assure the smooth international trade of equally treated products compromise the ability of regional organizations to institute effective management schemes and assure resource conservation.

The threats the international focus of the WTO pose for RFOs are an example of what action on fisheries subsidies at the WTO could mean for developing country national governance. Governance of fish stocks currently rests at the national level-since the UN Law of the Sea Convention of 1982 states have had sovereignty over stocks that lie within 200 nautical miles of their shores. As the subsidy reform issue highlights, management is a key aspect of development and resource security in developing coastal nations. National governance is thus an unavoidable part of a solution and holds many opportunities for conservation of fisheries resources by intervening in the domestic situation and creating management solutions. The government of Senegal, for example, announced in September 2003 the creation of four marine protected areas that are of vital importance to local fishers and the domestic economy. While this action was hardly a subsidy, at the level of the WTO some states may argue that any intervention by a government in the fish sector destroys the market. Hugo Cameron of the ICTSD gave a current example of this argument when he explained how, although many states are trying to use subsidies to reduce capacity by using subsidy funds to buy-back vessels or employ people to repair ports, states such as those in the "Friends of the Fish" may argue that this still provides people an incentive to remain in the industry and distorts the market.

As Carl Gustav Lundin warned, it can be a "slippery slope" when international bodies limit the interventions of national governments. In order to achieve sustainable development objectives, WTO rulings on fisheries must not interfere with developing nations abilities to use government support measures in their domestic economies and resource management sectors. Although governance of fish stocks rests at the national level, sustainable development objectives require that livelihood and food security be met at the local level, and that future solutions do not stop at questions of national access or resource use. Conservation, fair economic activity, and development are meant to be non-national goals, even though, as the fisheries subsidies issue highlights, they have often been dealt with in terms of national policy. Ending the analysis at the national level, though, is a fault. Current threats in the fisheries sector at the local level come from the way in which trade liberalization through the WTO may not assure that the distribution of benefits from export revenues correlate to any measure of development objectives. Increases in national export revenue do not necessarily represent changes in people's access to resources for food or livelihood. Instead, they may simply reinforce the existing economic structure within the community.

This is especially so in this case due to the lack of effective management of fisheries resources, which, as described, may only add to resource depletion under trade liberalization. Given the way the WTO does not monitor such local dimensions, it is incapable of assuring any outcome outside that which relates to the state as a single economic entity. A real resolution to the crisis would take the inner view of the national situation-most importantly here the level of resource management-into account and tailor policy solutions to this.

Conclusion

In order to achieve a true 'win-win-win' solution, subsidy reform must deliver not only benefits on the international level; global levelling of trade advantages in the economic sphere, net conservation of resources in the environmental sphere, and a targeting of resources toward development aims, but gains at all levels. This is perhaps the most challenging aspect of the role proposed for the WTO, a fundamentally international organization. In order to be successful, it needs to prove that it can improve the lives of people down to the local level. The concepts of sustainable development are an entry point into such progress, but, as this paper describes, there are serious limitations to relying on an international economic institution to deliver such goals. The threats which have been outlined; the lack of assurance that gains at the national level through increased export revenues and trade liberalization will deliver resource or livelihood security, the threat to national autonomy due to WTO intervention in the responses of state governments to the fisheries crisis, and the dire need for effective management in the fisheries of developing countries, represent needs on which WTO action as it is presently seen cannot fully deliver. Only when proposed reforms can assure that objectives are met in the three pillars of sustainable development-environment, trade, and development-as well as at all levels of organization and governance can any outcome be considered to be winning.

2

FISH CULTURE: RECENT DEVELOPMENTS

INTRODUCTION

The Government of India, assisted by the United Nations Development Programme and the Food and Agriculture Organization of the United Nations, are engaged in the Intensification of Freshwater Fish Culture and Training Project (FI:DP/IND/75/031) at the Freshwater Aquaculture Research and Training Centre (FARTC) of the Central Inland Fisheries Research Institute (CIFRI) at Dhauli, Bhubaneswar, Orissa. As part of the operation of the project, FAO assigned Dr J. Bakos as a Fish Culture Genetics Consultant for three months from 29 October 1980 with the following terms of reference:

1. To assist in establishing the fish genetical laboratory and in elaboration of studies on the selection of fish strains with better growth potential, meat quality, resistance to disease and longer breeding period;
2. To assist in the development of effective gynogenesis technique for carps;
3. To assist in identification and maintenance of land races of carps towards development of genetic pool.

During the assignment the consultant had detailed discussions with Dr V.R.P. Sinha, Project Director and Head of FARTC; Mr R.D. Chakrabarty, Head of the CIFRI Substation, Cuttack, and other scientists involved in breeding and genetical work. The Consultant wishes to thank all of them for their fruitful discussion. Thanks are also due to Dr Maharana of Cuttack Medical College for providing the cobalt 60 gamma ray instrument for the work.

Fish Species Cultivated in India

The following fish species are commonly cultured in India and are thus the most important from the point of view of genetic selection and hybridization.

Indian major carps:

Catla (Catla catla H.)

Rohu (Labeo rohita H.)

Mrigal (Cirrhinus mrigala H.)

Exotic carps:

Silver carp (Hypophthalmichthys molitrix Val.)

Grass carp (Ctenopharyngodon idella Val.)

Common carp (Cyprinus carpio L.)

Other fish species which are important for fish culture are as follows:

Indian minor carps:

Reba (Cirrhinus reba H.)

Bata (Labeo bata H.)

Calbasu (Labeo calbasu H.)

White carp (Cirrhinus cirohosa Bloch)

Cauvery carp (Labeo kontius Fedon)

The Indian major carps are the endemic species in India and their contribution to the total fish production is considerable. Thus, it is essential to improve their genetical capabilities and also to maintain their gene bank. Equally important are the Chinese carps, which also need proper genetical selection.

SUMMARY DESCRIPTION OF WORK UNDERTAKEN

Establishment of the Genetical Laboratory

The Genetical Laboratory consists of two rooms, one of them air-conditioned for under-taking biochemical investigations, such as electrofocussing, etc. The other room is for chromosomal investigations, gynogenesis and other genetic experiments. Another room will serve as a laboratory for selection and hybridization work. A fourth room will have provision for aquaria for rearing and keeping the progeny after gynogenesis and sex reversion. Here necessary equipment will be kept for marking, measuring and dissecting the fish. This room needs some drainage facilities for about 15 fibreglass tanks of different sizes.

After carefully examining the requirements of the genetical laboratory and the equipment already available, listed below is the additional equipment needed for the laboratory:

(1)	Gel electrophoresis apparatus (both disc and slab)	1
(2)	Densitometer	1
(3)	Spectrophotometer	1
(4)	Microscope, Dialux 20	2
(5)	Centrifuge, 500-5 000 rev/min	1
(6)	Centrifuge, refrigerated 10 000-13 000 g	1
(7)	Camera	1
(8)	Dissolved oxygen meter, electronic	1
(9)	Microscope, Olympus	1
(10)	Microscope, Stereo WILOM3	1
(11)	Refrigerator 265 1	1
(12)	Ultra-violet germicidal tube, 30 W Philips with extra tube	1+1
(13)	Refrigerator-deepfreezer 100 1	1
(14)	Refrigerator-deepfreezer 500-1 000 (-20°C)	1
(15)	Meat mincer	2
(16)	Electronic mixer	2
(17)	Fibreglass tanks 1 000 1, 500 1, 200 1 capacity	10 each
(18)	Electric aerator	5
(19)	Balance 0.5 g-1 000 g	1
(20)	Dia slide projector	1
(21)	pH meter, electrical Philips	1

Pond Requirements

Five different land races of catla, rohu, mrigal, silver carp, grass carp and common carp have to be maintained separately using a polycultural stocking system. As a safety measure, some specimens from all fish species must be maintained in two ponds in mixed conditions and permanently marked. This work requires 32 ponds (60 m^2).

The fish fry produced with artificial propagation should be reared separately. The expected number of different populations yearly (three species with five various land races need 15 ponds of 150 m^2), one intraspecific hybridization with polyallel crossings of five different females and males would need 25 ponds of 150 m^2 and, for intergeneric hybridization, 5 ponds of 150 m^2. Thus, altogether 45 ponds, 150 m^2 each would be required for fry/fingerling rearing. The area of testing ponds should be 2 000-5 000 m^2 each. About 30 such ponds will be required for undertaking experiments with land races, hybrid populations, etc. Further, about 10 ponds, 60 m^2 each may be required for other genetical research. Therefore the total requirements of the ponds would be:

1. 42 ponds of 60 m^2
2. 45 ponds of 150 m^2
3. 30 ponds of 2 000-5 000 m^2

Development of Gynogenesis Technique for Carps

Because the major carps and Chinese carps breed only during the monsoon season, common carp was selected to be the test fish (common carp also breed during December-January) and carefully selected males and females were induced to breed with pituitary hormone injections. The sperm was irradiated with cobalt 60 and also some by ultra-violet ray. Stickness of the eggs was removed by urea and salt solution. During the embryonic development of eggs, a proper check was made of the development of the gynogenetic progeny. Identification was made of haploid and diploid embryos. In certain countries a cobalt 60 gamma ray source in 100 000 Rad dose is given for irradiating the sperm. However, because of the difficulties in cooling the sperm here and also the duration of the cobalt 60 gamma ray, experiments were conducted to irradiate the sperm with the help of an available ultra-violet lamp. Appropriate duration of irradiation, intensity of the exposure and dilution of sperm were not known, so experiments were carried out to determine these. It was found out that the irradiation time should be 60 min, distance between the lamp and the sperm to be 12 cm, dilution 1:2 (dilution with frog Ringer) and the thickness of layer 2 mm. Mechanical shaking was used during the time of irradiation with a frequency of 120/min. Further experiments are necessary to find out the suitable method for cooling of sperm to 3-4°C during the irradiation.

The rearing of gynogenetic common carp progenies was carried out in plastic pools. They were initially given zooplankton collected from fish ponds for 28 days. After that they were given balanced artificial food consisting of sheep liver and heart and broiler chicken feed supplemented with vitamin A, B, C, D and E. The fry showed better growth rate than the normal fry of common carp. In order to have sex reversal in these fry, they have been fed with feed containing methyltestosteron. The result of the investigation will be known in due time. The above technique of gynogenesis has been carried out in collaboration with the counterpart scientists assigned to the consultant. Thus, these scientists are fully conversant with and trained to utilize this technique in other carps during their breeding season.

Identification and Maintenance of Land Races of Carps

Collection of various land races of cultivated carp species (preferably at the fingerling stage) from different water systems of India is in progress, the first consignments of fish populations was to arrive in January 1981. Those carp species are also to be collected from Cuttack. Together with the Cuttack strain, five different land races of each species are suggested to be collected. FAO, in collaboration with the United Nations Environment Programme (UNEP), expects to develop a programme to preserve the genetic resources of aquatic organisms. Therefore, it would be better to have a link of the collection and maintenance of land races of carp along with the above programme. After collection and arrival of fingerlings they need to be stocked in separate ponds. The various land races, as basic populations for research, have to be marked with fin clipping and operculum tagging. They should be reared in optimum conditions so as to achieve their best growth rate and early maturity. It is necessary to investigate the morphological/ biochemical characteristics of different populations.

CURRENT RESEARCH PROJECTS FOR FISH GENETICS

Selective Breeding and Hybridization

Studies on cytogenetical features of carp hybrids improving the strains of Indian major carps with crossing 1977-82.

Breeding of selected stock of grass and silver carps. Improved the stocks of Chinese carps 1977-81.

Studies on the morphology, growth and maturity of the hybrid between grass carp female and silver carp male 1978.

Hybridization between Labeo rohita × Cyprinus carpio; C. mrigala × C. Carpio; C. catla × C. carpio and C. catla × H. molitrix 1977-80.

In the field of genetics CIPRI attempted some preliminary experiments in interspecific and intergeneric hybridization, mainly for the purpose of improving the market quality of the fish (smaller head and deeper body, as in crosses between catla and rohu) and obtaining faster growth rates. Even though the results achieved are encouraging, it has not yet been possible to carry out adequate progeny testing due to a lack of facilities at the Cuttack farm. Genetic selection experiments to prevent inbreeding depression or to obtain strains with greater resistance to temperature variations common in tropical ponds, resistance to diseases and enhanced growth rates have not been attempted.

RECOMMENDED RESEARCH PROGRAMME AND ITS POSSIBILITIES

Methods and Phases of Research Work

Comparative Testing of Fish Species and Different Lines

The aim of comparative testing is to determine and compare the most important characteristics which are closely correlated with the productivity of species. The different phases of comparative testing are as follows:

1. The groups for testing have to be produced with artificial propagation. One female and one male can be used which are chosen by phenotypical selection with the best growth rate and appearance.
2. The groups of progenies will be reared in separate ponds and after six months they need to be stocked together in a common pond with group marking (150-200 fish from each population at least). The stocking rate of testing ponds must be the same as in commercial fish farming/intensive polyculture. When the number of stocked fish from one population are less than 100 fish, the testing work has to be repeated three to four times with the same groups of fish.
3. It is necessary to use organic and inorganic fertilizers to fertilize the water and also to feed the fish.
4. The fish population reaching a marketable size need to be harvested. Proper measurement of every individual is required for evaluating different characteristics.
5. The qualitative characteristics, such as the colour, scales and the formation of fins are not closely correlated with the productivity, therefore the evaluation of those only for the purpose of morphological characterization of the populations has any importance.
6. From the quantitative characteristics one has to evaluate the viability, growth rate, food conversion, quantity of meat, quality of meat (protein and fat content) and the reaction to fish diseases. Later, by producing inbred lines, the fecundity and the oxygen demand tolerance also has to be examined.
7. The results obtained need to be verified with variance analysis.
8. The productivity of various groups can be compared through a multifactorial, "100 score" system by the four most important characteristics (like viability, growth rate, food conversion, quality of meat).
9. If the test is arranged in more ponds with replicates, standard population control has to be used in all ponds,

which gives the possibility to compare the differences of environment among the ponds used. The common carp is suggested as a standard control considering its omnivorous feeding habits. The standard control can be produced in every year with a panmictic propagation, maintaining their quality and characteristics on the same level during a long time.

10. The results of pond testing experiments can be compared with the short laboratory tests and with the results of oxygen demand toleration (ODT) tests. The average of the above three methods gives a reliable final result of comparative testings.
11. The outstanding populations showing best characteristics will be used for the further selection and hybridization work.

Progeny Testing for Determining the Individual Breeding Value

The progeny testing is a suitable method to determine the individual breeding value of males and females after the productivity of their progenies.

The phases of the work are the following:

1. For the evaluation of males, the eggs of two females have to be fertilized with the sperm of chosen males separately, producing maternal half sister groups. Five or six males can be examined at the same time in their ten or twelve progeny groups from the two different females. For the evaluation of females one must follow the same system, using two different males.
2. The productivity of half sister groups are examined and compared by using the above mentioned comparative testing method (i.e., 4.1.1).
3. The average productivity of two groups originating from different females but from the same male shows the individual breeding value of the given male fish.
4. The best males and females in breeding value can be used for the intraspecific hybridization, or for improving the species in their characteristics.

Intraspecific Hybridization

The aim of intraspecific hybridization is to produce more productive hybrids by crossing different genotypes of females and males having the best results during the previous progeny testing. The phases of intraspecific hybridization are as follows:

1. Selection of various genotypes of the same species;
2. Inbreeding of the selected lines;
3. Polyallel crossing of the different lines;
4. Comparative testing of the F1 hybrid generation;
5. Propagation of the best hybrid producing parental lines for the fish seed production farms;
6. Producing three-line hybrids;
7. Producing four-line hybrids.

After the inbreeding of parental lines, the heterosis effect can be increased. The fastest method of inbreeding is the artificial gynogenesis. As a result of intraspecific hybridization, the increasing productivity of fish species of about 15-25 percent can be expected.

Artificial Gynogenesis and Hormonal Sex Reversion

The main purpose of gynogenesis is to produce inbred individuals having only maternal chromosomes for the further genetic work. For causing gynogenesis artificially, inactivated sperm can be used for fertilization of eggs, with a cold shock treatment after fertilization. For the inactivation of sperm a cobalt 60 gamma ray source or ultra-violet germicide lamp has to be used. The individuals of gynogenetic populations are females with diploid chromosome sets similar to the mother. The technique of artificial gynogenesis was demonstrated on the common carp. The method can be adopted for the other fish species with necessary modifications. For producing gynogenetic inbred male populations, methyl-testosteron hormone feeding has to be provided for the females at a young age to influence the sex formation in the direction of the male. The technique of sex reversion was demonstrated with gynogenetic common carp. The same method can be used for all fish species which are able to feed with artificial food.

The inbreeding coefficient value (F) of the first gynogenetic population is already 0.90, which value can be reached only after 12 generations of sister-brother pairing. For the purpose of increasing the inbreeding level, the gynogenesis can be applied from generation to generation. It is important to make a proper selection within the gynogenetic population to avoid the undesirable effect of inbreeding depression. The keeping and maintenance of gynogenetic inbred females and males requires more care than the normal ones because of their high breeding value and more sensitiveness to the environmental effects. Therefore, the location, rearing and keeping of this fish in small ponds is necessary, with a good water supply and other necessary conditions and to protect them from poaching. Some outstanding individuals can be kept in an indoor raceway system. The gynogenetic inbred populations or individuals can be used for improving the species with their useful characteristics, to make inbred lines for intraspecific hybridization, for producing monosex fish populations and also for genetic examinations for particular purposes.

Application of Gynogenesis for Improving Productivity of Fish Species

Producing and maintaining inbred lines for intraspecific hybridization: The greatest advantage of producing inbred lines is to obtain higher heterosis effect after the crossing of gynogenetic female and male lines.

The main phases of the work are as follows:

1. To select the best females for performance and morphological features.
2. Producing the first gynogenetic female generation (G1)
3. Rearing the gynogenetic fry suitably for producing sexually mature fish as soon as possible.
4. Selection by phenotype for the best growing and correct body formation.
5. Producing the second gynogenetic female generation (G2).
6. After dividing the G2 population into two parts, one part will be reared as a female group and the other part

has to be oversexed for producing gynogenetic males (G2).

7. Rearing the G2? and G2 groups suitably.
8. Brother-sister pairing can be used for producing the third gynogenetic generation (G2? × G2-G3?). The purpose of brother-sister pairing is to avoid the chance injuring of chromosomes.
9. Sex reversion in the G3? progeny group.
10. Brother-sister pairing for producing fourth gynogenetic generation (G2? × G3-G4)

This scheme can be continued by the previous pattern. During all the phases the best female or male are selected according to their better constitution. The inbred population produced can be used after phase five for intraspecific hybridization. After top crossing, good heterosis effect can be expected.

Producing semi-gynogenetic lines: The purpose of this work is to produce gynogenetic inbred male lines in cases when the hormone feeding is different, such as the case of silver carp and catla. The phases of work are as follows:

1. From population "A" selection to be made of the best females.
2. Producing the first gynogenetic generation (AG1?).
3. Crossing the AG1? female with a normal A? male. The result obtained will be a semi-gynogenetic bisexual population (50% ? and 50% ?).
4. Selection for the best females and males.
5. Producing the second gynogenetic generation (AG2?)
6. Crossing of the AG2? female with the normal A? male. The scheme can be continued by the previous pattern. After the third phase the gene pool of population becomes homozygotic to about 60 percent.

Producing monosex fish populations with gynogenesis: The purpose of this work is:

(i) to produce monosex female populations for pond fish culture because of their faster growth than that of the bisexual group of the same age;

(ii) the stocking of monosex populations in natural waters like reservoirs or undrainable lakes;

(iii) to develop inbred gynogenetic male lines is necessary for maintenance only in commercial fish producing farms or for fish seed production farms and pairing them with normal females;

(iv) after solving the problem of deep freezing the sperm, one can collect and store the sperm of inbred gynogenetic suitable males, so as to use it for producing monosex populations in other farms.

The phases of the work are as follows:

1. To select female from a male line
2. Producing the first gynogenetic population (G1 ?)
3. Producing a gynogenetic male population with hormonal sex reversion (G1)
4. Using the selected males for top-crossing by pairing them with normal females.

The progenies will be all females which may show improved characteristics.

Intergeneric Hybridization

Intergeneric hybridization consists of crossing fish belonging to different species. The aim of producing intergeneric hybrids is as follows:

1. To unite the useful characteristics of two different species in one of the new forms.
2. Producing a new construction in consequence of the heterosis effect.

Using intergeneric hybridization, the following results can be expected:

1. New construction with new feeding habits for composite fish culture or for water reservoirs
2. Better growth rates and survival rates
3. More restful behaviour, having better tolerance for reduced levels of oxygen (silver carp × bighead)
4. Better quality of meat (common carp × silver carp; rohu × catla)

5. Monosex triploid grass feeder fish (grass carp × bighead)
6. Better tolerance of water salinity for brackishwater fish culture.

In the field of intergeneric hybridization the scientists of India have achieved good results.

The phases of intergeneric hybridization are the following:

1. Studying the fish species characteristics; behaviour and genetic structures
2. Analysis of chromosome sets and karyotypes
3. Crossing the species, using artificial propagation
4. Rearing the hybrids in the most suitable environment for their expected habitat
5. Comparative testing of hybrids and the parental forms
6. Selection of the most suitable individuals
7. Artificial propagation experiments on the F1 hybrids
8. Stabilizing the new characteristics with inbreeding or back-crossing with the closer parental species which are similar in their main characteristics to the new hybrids
9. Describing the new hybrids' morphological, biological and economic characteristics.

Other Research

Biochemical polymorphysm investigations: These investigations are aimed at finding and using genetic markers for characterization and identification of fish populations and individuals. Such markers are the enzymes and proteins, including transferrins. The method is very important for characterization of species, land races and hybrids. Also this is useful in testing the success of gynogenesis in fully scaly fish species. The instrument for this method is the polyacrilamid gel-electrophoretic equipment which was ordered.

Chromosome investigations: These investigations have the aim of characterization of fish species and hybrids to study the possibilities of intergeneric hybridization. The method of chromosome investigation is well known by the scientists but some technical details have to be improved. These investigations will be also important in the future research programme with the aim of polyploidization.

List of Research Subjects Recommended for the Next Ten Years

1. To develop practical methods of artificial gynogenesis and hormonal sex reversion of catla, rohu and mrigal to produce highly inbred parental lines. Intraspecific hybridization of catla, rohu and mrigal, with crossing gynogenetically inbred female and male lines, to compare the productivity of the different populations.
2. To develop practical methods of artificial gynogenesis and hormonal sex reversal of silver carp and grass carp to produce highly inbred parental lines. Intraspecific hybridization of silver carp and grass carp with crossing gynogenetically inbred female and male lines. To compare the productivity of the different populations.
3. To develop practical methods of artificial gynogenesis and hormonal sex reversion of common carp to produce highly inbred parental lines. Intraspecific hybridization of common carp with crossing gynogenetically inbred female and male lines.
4. Fish chromosome investigations on Indian major carps and cultivated exotic carp species.
5. Biochemical polymorphism investigations on Indian major carps and the cultured exotic carp species to determine the transferrin and isoenzyme polymorphysm of different genotypes with a view to selection and hybridization of cultivated species. Intergenetic hybridization among:
   a. Indian major carps and minor carps
   b. Exotic carps
   c. Indian major carps and exotic carps.
6. To develop practical methods of producing polyploid and androgenetic fish populations in Indian major carps and the common carps.

7. Establishment and maintenance of the living gene pool of Indian major carps, collecting the different land races, elaborating practical marking systems, propagating for renewal of populations every five years and using them in the genetic and selection work.
8. Spreading the new fish hybrids into the commercial fish farms and fish seed producing farms in the form of inbred female and male lines.

EXPECTED RESULTS OF THE PROJECT RESEARCH

Some results of project research will provide a scientific basis for future research but other results should assist the producing sector to develop better results in actual fish farming. As a result of chromosome and biochemical investigations one can become acquainted with the genetic structure of cultivated fish species and obtain additional information to determine the further directions of the work of selection and hybridization.

By the comparative testings and progeny testings the best populations and individuals are selected in terms of their productivity.

Using the best strains, the population can be improved, not only in the Institute but also in the fish farms. By improving productivity of species with intraspecific hybridization, on the basis of Hungarian experience, the following results can be expected: in the survival rate + 20%, growth rate + 20%, conversion of food + 15% and increasing percentage of eatable meat + 2-3%.

The monosex female populations provide 10-15% higher growth than the bisexual mixed ones in the same environment. By artificial gynogenesis and hormonal sex reversion inbred female and male lines can be produced, and from their heterosis hybrids 30-40% higher productivity can be expected. By improving the fish species by the above mentioned methods, the productivity of composite fish culture should be increased about 15-20% without any extra financial investment.

3

MITIGATION OF GENE FLOW PROBLEM

By contrasting the pattern of genetic structure at putatively neutral microsatellites with the pattern at the heat-shock cognate protein (Hsc70), Hemmer-Hansen and colleagues convincingly show the existence of adaptive divergence despite high gene flow in the European flounder Platichthys flesus. This result is unexpected because if gene flow is high it should preclude local adaptations by rapidly blurring allelic frequency differences between populations.

If gene flow is the dominant force controlling both, the pattern of genetic structure inferred with neutral markers is thus expected to coincide with the pattern at loci under selection (Conover et al., 2006). This relationship provides the basis for using information from population genetic studies with neutral markers in the context of conservation of marine resources and fisheries management.

The Hemmer-Hansen study now clearly shows not only that local adaptation is possible in the sea, but also that patterns of differentiation at selected and neutral markers can be strikingly different. This means that current management practice could misidentify management units, leading to erosion of genetic resources.

This study comes at a time of exciting research on the population genetics of marine organisms. With the increased application of molecular tools, many commonly accepted perceptions have been challenged in the field. Starting from the general belief that populations of marine organisms, being large and inhabiting an environment with no obvious barriers to genetic flow, are essentially homogeneous with at most a pattern of isolation by distance shaped by the duration of larval dispersal, we are now facing a much more complex picture.

The wide application of microsatellites allows us to easily measure the connectivity between marine populations, resulting in the detection of significant albeit weak genetic differentiation. It also allows us to estimate the importance of larval retention in local areas, questioning the idea that dispersal is determined by the extent of larval duration only.

In addition, the finding of a very small effective population size in many abundant species and the existence of trends of loss of genetic variation over time, disputes the idea that these species are characterized by large population sizes and thus are not prone to genetic erosion.

Finally, the occurrence of rapid contemporary evolution mediated both by environmental and human-related selective pressures, and the emerging body of evidence that indicates the importance of behaviour at the larval and adult stage in shaping the structure of marine populations, identifies a complicated situation.

In this context, using a candidate gene approach, Hemmer-Hansen et al. (2007) compared the pattern of genetic variation at eleven microsatellites with that at one intron of Hsc70, a gene involved in the response to thermal/osmotic stress and pollution.

Analysing population samples from the north-eastern Atlantic, the authors found a fivefold higher genetic differentiation, measured by FST, at Hsc70 than at the microsatellites.

When the confidence interval of single locus divergence was reconstructed under the neutrality hypothesis, Hsc70 was identified as a strong outlier, indicating the existence of directional selection. Notably, the result of higher differentiation at Hsc70 is even stronger after the standardization of genetic divergence, taking into account the maximum divergence attainable at each locus. With this approach, the authors corrected a possible bias resulting from microsatellites' hypervariability, because maximum divergence is limited by the observed homozygosity of the sub-populations (Hedrick, 2005).

In this respect, this study augments similar results recently obtained for another putatively selected locus, the pantophysin locus (PanI), in the walleye pollock (Canino et al., 2005), by showing that the observed higher divergence at functional genes is not due to a comparison between markers with different mutational properties.

Importantly, Hemmer-Hansen et al. (2007) also found strong differences in the pattern of genetic structure between neutral and selected loci.

Thus, the results of this study differ from that obtained with the PanI locus in cod around Iceland (Pampoulie et al., 2006), where a standardized higher differentiation was found at PanI with respect to microsatellites, but the pattern of differentiation was congruent between the two sets of loci. In that case, the authors concluded that oceanographic and environmental factors acted together in shaping genetic diversity in that species.

In the European flounder instead, comparisons of samples from the western and central Baltic with those of the North Atlantic detected strong differences: neutral markers suggest homogeneity between the three locations, while Hsc70 reveal a sharp break in allelic frequencies. This pattern is striking because of the temperature/salinity gradient occurring in the Baltic Sea, which could represent the selective force acting on Hsc70.

The exact evolutionary and conservation implications of the Hemmer-Hansen results are still unresolved and depend on the understanding if selection pressures occurred in the past and the populations have not yet attained equilibrium, or if selection is currently taking place. This question could be answered by comparing the performance of different genotypes in different salinity/temperature conditions in common garden experiments that, of course, can be difficult with an organism like flounder. At present, Hemmer-Hansen et al. (2007) argue that migration-drift equilibrium has been reached at neutral markers in the investigated samples, citing both the observed strong differences in microsatellites among several samples and comparisons with other species in the same environmental setting.

As allelic frequencies tend to approach equilibrium much faster in loci under selection than in neutral loci, the message is that the higher differentiation at Hsc70 is maintained by contemporary selection.

In this case, it is important to recall that, at equilibrium, FST is inversely related to Nm, the product of effective population size N for the proportion m of immigrants. Therefore, estimates of effective population size and application of realistic models of gene flow are needed to estimate m and quantify the genetic cost of immigration for the populations and its possibility to preclude local optimal adaptation.

In this regard, it is noticeable that the Baltic populations, on which the apparent selection takes place, are not fixed for a specific allele of Hsc70. Then, if populations are at equilibrium, a cost in terms of mortality linked to gene flow may be present, and, considering that at the genome wide level many loci can be under selection, the genetic load experienced by the population might be extremely high. Thus, if the existence of local adaptation is generally confirmed, an important link between population genetics and dynamics can be foreseen, and the application of both selected and neutral markers will be needed to effectively resolve the stock structure of marine fish.

THE BIOCHEMISTRY AND MOLECULAR GENETICS OF PROSTATE CANCER PROGRESSION

The research in this laboratory focuses on elucidating the biochemical pathways and molecular genetic mechanisms associated with the development and progression of human prostate cancer. Within the United States, there has been an increase in both prostate cancer incidence and mortality over the last several decades. The etiology of this disease is unclear, and most prostate cancer patients have no known risk factors for either disease development or rate of progression. It is known that African-Americans have a two-fold greater incidence of prostate cancer than Caucasians. In general, prostate carcinoma is a slow growing malignancy that is diagnosed by the physician in a variety of clinical stages. These range from indolent disease, which never manifests itself clinically and is only apparent upon autopsy, to advanced metastatic prostate cancer, which can be widely disseminated and extremely difficult to treat.

The biochemical and molecular events associated with this pattern of progression are currently very poorly understood. We are conducting studies in three areas to elucidate some of these mechanisms.

The first is to verify recently observed alterations in Y chromosome number in prostate tumors, and to determine the biological and clinical significance of this event. The second is to investigate Clar1, a novel gene cloned in our laboratory that exhibits enhanced expression in more advanced prostate cancer, and to understand its biochemical function and role in disease progression.

The third area consists of studies examining the role of the insulin-like growth factors (IGFs), their receptors (IGFRs), and the insulin-like growth factor binding proteins (IGFBPs), in the normal and malignant growth of prostate tissues and in apoptotic mechanisms.

The Y chromosome has been shown to be either lost or gained in human prostate tumors; however, the biological and clinical significance of Y loss in prostate cancer remains to be determined. One contention is that Y minus prostate tumors exhibit a more poorly differentiated phenotype and tend to predict a less favorable clinical outcome. Several studies have shown this loss to be specifically associated with malignant epithelium in the prostate and that Y loss can occur in cases of prostatic intraepithelial neoplasia, suggesting it may be an earlier event in prostate tumor development than previously thought.

However, other studies do not detect loss of Y and still others point to evidence for a gain of the Y chromosome in prostate tumors. It is of note that several genes encoding transcriptional activators reside on the Y chromosome.

It is conceivable that loss or gain of one or more of these genes could compromise downstream gene expression events resulting in alterations in the growth or metastatic capacities of prostate tumor cells.

We have demonstrated the differential expression of two chromosome Yencoded genes, sex-region Y (SRY) and zinc-finger Y (ZFY), in human prostate tumors. Lack of SRY and ZFY expression in a number of these tumors, nonetheless, did not correlate with loss of these genes as determined by Southern analysis.

To investigate whether gain or loss of Y occurs in prostate cancer and the relationship of this event to tumor progression, we are examining Y chromosome number in prostate tumor touch preparations from human prostate tumors of different stages and grades. We are currently using probes specific for chromosome Y and multicolor fluorescence in situ hybridization analysis to determine the frequency, location, and clinical significance of Y loss in this disease.

The results of our preliminary survey on 42 prostate tumor/ normal pairs suggest that alterations in Y chromosome number occur in only a very small subset of these tumors. Regional analysis on a larger set of samples is currently underway.

SRY is a chromosome Y-encoded gene that is essential to male determination in both humans and mice. It encodes a transcriptional activator that is a member of the high mobility group family of proteins. First thought to be expressed only in the male developing gonadal ridge and adult testis, SRY was shown by our group to be expressed in the human prostate, and has been subsequently demonstrated by others to be expressed in a variety of fetal and adult tissues.

We later demonstrated that SRY transcripts are expressed in some human prostate tumors, but not in others. We are currently developing antibodies to SRY to determine the expression levels of the SRY protein in freshly obtained human prostate tumors and in paraffin-embedded sections of tumor tissue.

These results will be correlated with Y chromosome status, tumor pathology and patient disease-free survival. We are also constructing an inducible vector system containing the SRY cDNA in order to test the effects of SRY expression on the growth and tumorigenesis of the PC-3 prostate cancer cell.

MOLECULAR MECHANISMS OF PROSTATE TUMOR PROGRESSION. WINTER, CVETKOVIC, TRICOLI, in collaboration with GOLEMIS

The molecular events that are involved in prostate tumor progression are largely unknown. An understanding of these events is essential to determining how tumors progress from an early, well differentiated, localized lesion to a poorly differentiated metastatic tumor.

Understanding these events may also provide clues as to why in some individuals early prostate cancer lesions can remain indolent and never manifest themselves clinically, while in others these lesions progress to clinically diagnosed prostate cancer. We are performing studies aimed at determining whether consistent gene expression events occur during the progression of prostate tumors from pathologic stage B to stage D disease.

By utilizing differential display methodologies and RNA isolated from prostate tumors of different pathologic stages, we are working toward identifying gene expression differences that occur during tumor progression. Using this methodology, we identified a novel cDNA, that we have named Clar1, that is expressed to greater levels in prostate tumors of later pathologic stage and higher Gleason grade. We have also developed a quantitative reverse transcription (RT)-PCR assay that allows us to determine Clar1 transcript levels in small biopsy samples.

The full-length Clar1 cDNA is 2599 nucleotides in length and is novel except for a CpG island sequence near the 5' end. The Clar1 gene is differentially spliced resulting in two transcripts of 2.6 and 2.0 kilobases.

The deduced amino acid sequence consists of a proline-rich 33.8 kilodalton protein that has a number of src homology

(SH)3 binding consensus regions. These regions are associated with proteins that participate in signal transduction pathways. Clar1 also contains a consensus serine phosphorylation site near the carboxy-terminus. Taken together, these motifs suggest that Clar1 may participate in some aspect of signal transduction, perhaps as an adaptor or effector protein, and that its activity may be regulated by phosphorylation. We are currently studying the expression of Clar1 in prostate tumors using in situ hybridization, and raising antibodies to the protein for use in immunohistochemistry and western blot analysis. We also plan to search for possible binding partners for Clar1 using yeast two-hybrid methodologies to obtain clues about its function.

Furthermore, we are employing antisense strategies to inhibit Clar1 expression in the LNCaP prostate cancer cell line to determine the effect of turning off expression on growth and tumorigenesis. The overall aims of this work are to identify and characterize the role of Clar1 in events associated with the clinical and pathologic progression of prostate tumors.

THE INSULIN-LIKE GROWTH FACTORS AND PROSTATE CANCER. WINTER, TRICOLI, IN COLLABORATION WITH WATKINS-BRUNER

The IGFs, IGFRs and IGFBPs are important factors in prostate cancer cell growth. IGF-1R is essential for maintaining cellular sensitivity to apoptosis, and both IGF-1 and IGF-2 can influence prostate tumor growth and alter cellular calcium flux. IGF-1 and IGF-1R levels are known to decrease and IGF-2 levels increase as prostate cancer cells become more aggressive. IGFBP3, the major carrier for IGF-1 in human serum, is decreased in serum from patients with prostate cancer, while IGFBP2 is elevated.

Also of interest, IGFBP3 is positively regulated by the tumor suppressor p53 and is a target of prostate specific antigen (PSA), a protease that is elevated in the serum of prostate cancer patients. We are currently examining the levels of IGFBP2 and IGFBP3 in serum from prostate cancer patients, and in serum from men at greater risk of developing prostate cancer. These samples are being procured through the Prostate Risk Assessment Program (PRAP) recently established here at

Fox Chase. Our study results have demonstrated that African-American men have significantly lower mean plasma levels of IGFBP3 than Caucasian-American men with a P-value of 0.0045. African-American men have a two-to three-fold greater incidence of prostate cancer than Caucasian-Americans; these results support the hypothesis that lowered plasma IGFBP3 levels may lead to increased IGF-1 bioavailability, and thus greater stimulation of prostatic epithelium by IGF-1. Men in the PRAP program will be followed yearly to determine if alterations in the level of these binding proteins predict development of the disease, and how they might correlate with PSA levels and various demographic factors.

Studies using prostate cancer cell lines are also being conducted to determine the role of IGF-I and-II in prostate tumor growth and how the IGF-1 receptor and IGFBP3 levels affect the bioactivity of these proteins. The overall goal of this work is to determine what role the interactive IGF system might play in the growth and progression of prostate cancer, and to elucidate the biochemical mechanisms underlying the growth stimulation of prostate cells.

MUSCLE BIOPSY TECHNIQUE FOR ELECTROPHORESIS ANALYSIS OF FISH FROM THE GENUS BRYCON

Protein and mitochondrial DNA have been used as molecular markers to assess variability in stock identification studies of fishes. Protein and mtDNA used for electrophoretic analysis are extracted from tissues, which often leads to death of the individuals. In this study, we present a skeletal muscle biopsy procedure to extract mitochondrial DNA that does not require specimen sacrifice.

Eighty pirapitinga-do-sul (Brycon opalinus) were biopsied by the present technique, with no mortalities recorded. Total DNA was extracted from muscle and digested by restriction enzymes ApaI and HpaI. The mtDNA fragment patterns were hybridized with 32P-labeled pirapitinga-do-sul mtDNA probes. The described technique is simple and may be useful in protocols requiring tissue extraction for DNA and protein analyses without loss of the individual investigated.

The genus Brycon belongs to the subfamily Bryconinae and has about 60 species (Howes, 1992), some of which are distributed throughout Brazilian watersheds (Géry and Mahnert, 1992). Anthropogenic influences have caused a reduction in natural populations of some Brycon species. For instance, in the Vale do Paraíba watershed, hydroelectric power plant dams, pollution and overfishing have brought the economically important Brycon insiginis (Piabanha) close to extinction (Salgado and Chain, 1997). Rehabilitation of endangered species depends on the ability to characterize natural and captivity populations. Evaluation of genetic differences within and among populations is the first step to a successful restocking program (Toledo-Filho et al., 1992).

Biochemical and molecular tools for identification of proteins (Revaldaves et al., 1997), mitochondrial DNA (Danzmann et al., 1991) and nuclear DNA (Wright, 1993) polymorphisms have been developed to study genetic variation in different fish species. Population assessment or captivity stock discrimination using electrophoretic characterization requires a reliable source of biological material. The aim of this study was to establish a biopsy technique to obtain skeletal muscle samples from Brycon fish for mitochondrial DNA analysis without killing or seriously damaging the sampled individuals.

MATERIAL AND METHODS

The present muscle biopsy technique was developed at the "Estação de Hidrobiologia e Piscicultura da Hidroelétrica de Paraibuna-CESP" using 80 Pirapitinga do sul (Brycon opalinus) individuals from the aquaculture station.

Biopsy Method

Animals were anesthetized to equilibrium loss with benzocaine (ethyl-p-aminobenzoate) at a concentration of 1:5000 (Ross and Geddes, 1979). Scales were removed mid-way between the dorsal and adipose fins (Figure 1). Using a surgical blade, a longitudinal slice was removed from the dorsal region (Figure 2). Skeletal muscle tissue was then collected and stored in 95% ethanol at room temperature until use. The biopsied area was treated with methylene blue (1%) to prevent Saprolegnia growth.

After equilibrium recovery, specimens were transferred to small tanks (10 m3), where 0.5% methylene blue was continuously added to the inlet water for an additional two days.

Total DNA Extraction

Total DNA was extracted from the biopsied tissue according to the modified protocol of Taggart et al. (1992). Briefly, muscle was cut into small pieces and placed in 500 l STE buffer (0.1 M NaCl, 0.05 M Tris-HCl and 0.01 M Na2 EDTA, pH 8.0), 10 l SDS 10%, and 30 l proteinase K (10 mg/ml). The solution was left to digest for 2 h at 50oC. The tubes were inverted several times to accelerate the digestion process. After the digestion process was concluded, 3 l of DNAse free-RNAse was added, and the suspension was incubated for 30 min at 37oC. The resulting digestion mixture was extracted once with 500 l of buffer saturated phenol, pH 8, once with pheno-chloroform-isoamyl alcohol (24:1), and once with chloroform-isoamyl alcohol. The DNA was then precipitated in 1 ml cold ethanol and sodium acetate (3 M, NaOAc, pH 5.3) (Maniatis et al., 1982). DNA pellets were dried and dissolved in 100 l TE buffer (1 mM Tris-HCl, 7.6 and 0.1 mM EDTA, pH 8.0).

Mitochondrial DNA Detection

Two microliters of each total DNA extract was digested with restriction endonucleases ApaI (GGGCC/C) and HpaI (GTT/AAC). Digested DNA samples were run in 0.8% horizontal agarose gel (7.7 cm) for 1 h. Gels were denatured (0.5 M NaOH, 1.5 M NaCl) and neutralized (1.0 M Tris-HCl, 1.5 M NaCl, pH 7.2). Then, DNA fragments were immobilized on a nylon membrane (Biodyne Membrane, Gibco BRL) by Southern transfer (Southern, 1975).

Mitochondrial DNA was detected by hybridization with homologous probes of Brycon opalinus cloned at the Laboratório de Genética e Cardiologia Molecular, Instituto do Coração, INCOR. Briefly, mitochondrial DNA was isolated as described by Chapman and Powers (1984), then cut with the restriction enzyme BamHI.

The resulting fragments of 7.710, 4.930 and 2.125 base pairs were cloned into the Bluescript plasmid (Stratagene) cut with the same enzyme. The homologous mtDNA probes were labeled with [-32P]dCTP by random primer extension (Ready-To-Go DNA labeling Beads (-dCTP), Pharmacia) for hybridization with total DNA muscle extractions.

RESULTS AND DISCUSSION

The development of sampling strategies to obtain genetic markers from protein and/or DNA is an essential step for genetic studies. A number of procedures have been described to obtain skeletal muscle tissue for protein electrophoresis analysis in population studies, stock management, and genetic improvement (Uthe, 1971; Crawford et al., 1977; McAndrew, 1981; Leitner and Isely, 1994). Muscle is a very important source of isozyme systems, and approximately 60% of the loci routinely used in genetic analysis is expressed in this tissue (Morizot et al., 1990). The amount of muscle tissue required for isozyme detection in the laboratorial procedure is 1-3 g (Falk et al., 1996). The technique presently described yielded an average of 2.8 g of muscle tissue from specimens with 12 to 25 cm in length, which is suitable for isozyme analysis.

In the present study, 80 specimens underwent biopsy. After 3 weeks, there were no mortalities, and the specimens were returned to their typical shoal environment. The wound caused by the biopsy healed completely within 2 weeks (Figure 3) without tissue necrosis or other serious local damage to the individuals. mtDNA fragments present in the total DNA extraction were efficiently hybridized with the 32P-labeled homologous probes and visualized after 12-h X-ray film exposure at room temperature.

Mitochondrial DNA is an important genetic marker for phylogenetic and populational studies among different taxa (Avise and Lansman, 1983). mtDNA studies in fish have relied on soft-tissues such as liver, heart or kidney (Ferris and Berg, 1987) in which the donors are usually sacrificed for tissue removal.

Nonlethal samples for total DNA extraction have been obtained from blood (Billington and Hebert, 1990) or adipose fins (Hilsdorf, A., unpublished results). However, in this case, mtDNA is present at a lower concentration than in muscle tissue; hence, total DNA yields from blood or adipose fins are expected to be less than the nonlethal DNA extractions from muscle tissue presently described.

EMBRYONIC AND GENETIC MANIPULATION IN FISH

Fishes, the biggest and most diverse community in vertebrates are good experimental models for studies of cell and developmental biology by many favorable characteristics. Nuclear transplantation in fish has been thoroughly studied in China since 1960s. Fish nuclei of embryonic cells from different genera were transplanted into enucleated eggs generating nucleo-cytoplasmic hybrids of adults.

Most importantly, nuclei of cultured goldfish kidney cells had been reprogrammed in enucleated eggs to support embryogenesis and ontogenesis of a fertile fish. This was the first case of cloned fish with somatic cells. Based on the technique of microinjection, recombinant MThGH gene has been transferred into fish eggs and the first batch of transgenic fish were produced in 1984.

The behavior of foreign gene was characterized and the onset of the foreign gene replication occurred between the blastula to gastrula stages and random integration mainly occurred at later stages of embryogenesis. This eventually led to the transgenic mosaicism. The MThGH-transferred common carp enhanced growth rate by 2-4 times in the founder juveniles and doubled the body weight in the adults. The transgenic common carp were more efficient in utilizing dietary protein than the controls. An "all-fish" gene construct CAgcGH has been made by splicing the common carp-actin gene (CA) promoter onto the grass carp growth hormone gene (gcGH) coding sequence. The CAgcGH-transferred Yellow River Carp have also shown significantly fast-growth trait. Combination of techniques of fish cell culture, gene transformation with cultured cells and nuclear transplantation should be able to generate homogeneous strain of valuable transgenic fish to fulfil human requirement in 21st century.

4

NUCLEAR TRANSPLANTATION IN FISHES

INTRODUCTION

Lives on earth first appeared over 3 billion years ago. As a result of evolution, vertebrates are the utmost advanced life form distributing from seabed to high-mountain. Fishes, at the lower stage of evolution but being the biggest community in vertebrates, include about 21,700 to 28,000 species that take over almost half of the total number of vertebrates1. Fishes are also models for experimental study as the eggs are large, fertilization and development are externally and the embryos are transparent. In addition, mono-sexual breeding and crossbreeding between many distantly related species could be done in fishes.

Fish culture is one of the earliest activities of human civilization. Time dating back to 2500 years ago when China was at the Spring and Autumn Period, an ancient Chinese named Fan Li created a great literature "Handbook of Fish Culture" in which domestication and cultivation of common carp in ponds were described in detail. This is the first monograph of fish culture in the world. Since then, Chinese farmers enjoyed a long tradition of selective domestication resulting in very valuable "four farming species" of fish for pond culture. These are black carp (Mylopharyngodon poceus), grass carp (Ctnopharyngodon idellus), silver carp (Hypophthalmichihys molivrix) and big-head carp (Aristichthys nobilis). Based on the feeding habits of these fishes, Chinese farmers also invented "multi-culture method" of stocking the four farming species together in the same water body. This is in fact a sustainable ecological system by that farmers are able to gain maximum output with minimum cost. These domesticated species and the multi-culture method have been favorably adopted by other nations abroad.

In addition to the traditional experiences of fish farming, scientists in China initiated some fundamental research in fish cell and developmental biology and molecular genetics leading to a break-through in fish biotechnology. The major achievements are pointed as follows. In 1960's, Professor Tung T. C. first introduced the art of nuclear transplantation with fish to study the interaction between nuclcus and cytoplasm. Tung and his colleagues succeeded in generation of "nuclear-cytoplasmic hybridized fish" between different species. In early 1980's, the first somatic cell cloned fish was derived from kidney cells by nuclear transplantation. In 1984, the first batch of transgenic fish was also generated. The development and combination of technologies of nuclear transplantation and gene transfer in fish predicate that a new era in fish breeding is coming; fish directional breeding will revolutionize the traditional fish farming industry.

NUCLEAR TRANSPLANTATION IN FISH

The concept of nuclear transplantation dates back to 1938 when a German biologist, Hans Spermann, proposed an experiment that would evaluate the relative importance of the nucleus and cytoplasm in controlling early developmental events. In the early 1950's, the technology of nuclear transplantation was first demonstrated with frogs by Briggs and King and amphibians were becoming most commonly used for studying nuclear transplantation. These studies revealed that the nuclei undergo restriction in developmental capacity as cells became differentiated. Although several experiments suggested that differentiated somatic cells still had developmental totipotency, the cell differentiation didn't necessitate any irreversible changes in nuclear genetic material. Some subsequent experiments with differentiated somatic cells from Xenopus and Ranna pipens could hardly supported the previous cases because that nuclei from these cells only supported the ontogenesis to feeding-stage of tadpole but not to an adulthood.

Thus, whether ultimately differentiated cells have the capacity to support the reconstructed embryos to complete their full ontogenesis still remains unanswered. Illmensee and Hoppe in 1981 first succeeded in nuclear transplantation in mammals. They reported that an inner cell mass nucleus could support early development of an enucleated egg in mouse. Three years latter, McGrath and Solter reported nuclear transplantation in mice by using micrioinjection coupled with cell fusion technique.

Since then, embryonic nuclear transplantation in sheep, calve, rabbit, porcine and goat were successively done. In 1997, the birth of cloned lamb "Dolly" derived from an adult mammary gland cell marked the beginning of a "golden-age" in animal cloning. Recently, success has also been announced in cloning mice, goat, and cattle derived from differentiated cells: fetal fibroblasts, muscle cells, cumulus cells or oviductal cells. Successes on somatic cloning conclusively revealed that differentiated adult cells still remain totipotent and maintain the whole genome to support normal development to term.

In amphibian and mammals, nuclear transplantation has been successfully done within the same species. In other words, the donor of nucleus and the host of enucleated eggs must come from the same species. On the other hand, "Tung's fish" brought an absolutely new story. Tung's embryonic nuclear transplantation in fish involves three steps in general.

(1) *Preparation of donor cells*: the blastoderm was separated from the yolk with a fine glass needle, and carefully placed in Holtfreter's dissociation solution for further separating into individual cells.

(2) *Preparation of unfertilized host eggs*: the mature eggs stripped from the female were dichorionized with a pair of forceps, and enucleated by inserting a sharp glass needle into the egg cytoplasm just underneath the site of the second polar body.

(3) *Nuclear transplantation*: after donor cell was sucked into the micropipette by a slight negative pressure and microinjected into the enucleated eggs, the reconstructed embryos were put into Holtfreter's solution for further development. Great successes of fish nuclear transplantation were achieved between different genus (common carp (Cyprinus Carpio Linnaeus) and crucian carp (Carassius aurantus) and different subfamilies (grass carp (Ctenopharyngodon idellus) and blunt-snout bream (Megalobrama amblycephals)). The nuclear-cytoplasmic hybrid fish revealed that while most phenotypic characteristics were controlled by the nucleus, a few were controlled by the cytoplasm or by both.

In 1984, the first somatic cell cloned fish derived from short-term cultured kidney cells of triploid crucian carp was produced4. In this experiment two rounds of nuclear transfer were carried out. In the first round, the nucleus (3n) was transferred into the enucleated eggs (2n) of crucian carp. There were 41% of the injected eggs developed into blastulae but no future development occurred. Nuclei from the blastulae were taken for second round of transfer into another batch of enucleated eggs (2n). In the second round, 8 gastrula generated and one of them developed into a fertile female fish (1.2%) with normal morphological features of crucian carp. The chromosome number was triploid (3n=150). This was the first case of cloned animal with somatic cells in fishes. It suggested that some nuclei of somatic cells, following two rounds of nuclear transplantation, could be reprogrammed to totipotent status as zygotic nucleus does.

Nuclear transplantation in fishes, as an approach for studying the relative roles of nucleus and cytoplasm in controlling the characteristics, had also been considered as a helpful method in obtaining new farming strains of fishes. Yan3 indicated that the nuclear-cytoplasmic hybrid was 22% higher in growth rate, 3.8% higher in protein content and 5.58% lower in lipid content than that of the control, respectively. The art of nuclear transplantation in fish, however, not only requires very skilful manipulators but also bring about very low efficiency in producing the nuclear-cytoplasmic hybrids. The application of nuclear transfer fish in fishery is, thus, under restriction to a certain extent.

GENE TRANSFER IN FISH

People never give up their effort on pursuing of breeding new strains of farming species with high quality. Meanwhile, researchers are seeking for some genetic materials other than the whole cell nucleus for transfer. Around the late 1970's, researchers in China introduced the genomic DNA of common carp into the fertilized eggs of Mud carp (Cirrhina molitorella), a tropical species. About 8% of the "total DNA-transferred" founder Mud carp showed increase of cold-resistance.

In early 1980's, with the advancement of techniques in molecular cloning and embryonic micro-manipulation, recombinant genes are able to be constructed and transferred into the host animals. The transgenic "super mouse" was the most stimulating report in transgenic studies.

The first batch of faster growing transgenic fish was generated by introducing a recombinant human growth hormone (hGH) gene capped with a mouse metallothionein-1 (MT) gene promoter into goldfish in 19845. In this study, fertilized eggs of goldfish were obtained by artificial spawning and insemination and the chorion of eggs were removed by digestion in 0.25% trypsin solution. About 1-2nl of DNA solution containing about 105-106 copies of the MThGH gene was delivered with a micromanipulator into the germinal disc just underneath the second polar body. All the manipulated eggs were carefully put into Holtfreter's solution for further development. By Southern hybridization, more then 50% of the founders are transgenics.

Since then, dozens of laboratories all over the world began to show great zeal for the study of transgenic fish, and gene transfer into fish embryos were performed in several species, such as rainbow trout. (Salmo irdeus Gibbonsi), Atlantic salmon (Salmo salar Linnaeus), tilapia (Oreochromis nilotica), medaka (Oryzias latipes), common carp, zebrafish (Danio rerio), loach (Misgurnus anguillicaudatus), catfish (Parasilurus asoltus

Linnaeus), etc. Some other techniques, e.g. electroporation and sperm-mediating were also successfully employed in producing of transgenic fish. In addition to gene transfer with GH gene, other types of genes were also employed in gene transfer in fish. For example, GFP (green fluorescent protein) gene was transferred into fertilized eggs of zebrafish as a reporter and AFP (antifreeze proteins) gene was introduced into Atlantic salmon to gain freeze-resistant salmon. However, gh is so far the most commonly used and most thoroughly investigated type of transgenes.

It was in a transgenic fish model that the behavior of a foreign gene, this time the MThGH gene, in embryogenesis was intensively studied by Zhu et al. Southern blotting revealed that the behaviors of MThGH gene in host fish were a dynamic process, including replication, degradation, concatemer formation, and integration during embryogenesis. The replication began at very early stage of cleavage and suddenly took place at late-blastula to early-neurula stages. After neurula stage, most of the foreign gene were migrating with the host chromosomal DNA as revealed on the agarose gel by electrophoresis. It was suggested that foreign gene at these stages was in form of either large concatemers situated outside the chromosomes or integrated into the host genome. If the integration occurred at germ line cells of the host fish, the transgene could be inherited to their offspring. Northern hybridization showed that the transcripts of hGH gene could only be found at post-late-gastrula stage, which was consistent with the timing of the differentiation of fish embryonic cells.

As a result of the expression of MThGH gene, transgenic fish showed significantly faster growing trait. On the other hand, a certain proportion of the founders did not show growth enhancement and a few even grew slower than the control. This unexpected observation was reasonable when the multi-sited integration and transgenic mosaicism were taken into account. In the case of the generation of transgenic mammals, foreign gene had been microinjected into the zygotic pronucleus, which resulted in the integration of foreign gene occurring mostly before the first cleavage. It was believed that the status of transgene in founder mammals was in a homozygous form, which could be transmitted to next generation in a Mendelian manner. However, pronucleus of fish eggs is not visible and the foreign gene could only be microinjected into their cytoplasm.

The foreign gene integration in fish was found to be spanned over a long time course from gastrula to late developmental stages resulting in multi-sited integration and consequent transgenic mosaicism. Three categories of transgene integration could be deduced: functional integration, silent integration, and toxic integration. It is only in the functional integration that the transgene was integrated into host genomic sites suitable for expression and therefore showing "fast-growth" trait. Thus, the founder generation (P0) of transgenic fish was far from a genetically homogeneous strain.

The specific growth rate (SGR) of P0, F1, and F2 MThGH-transgenic fish, as well as of F4 generation was significantly higher than that of the controls. Bioenergetic analysis on MThGH-transgenic fish compared with controls was thoroughly worked out.

Compared with the controls, transgenic fish had a significantly higher proportion of food energy channelled to G and a significantly lower proportion of that channelled to R and U. The transgenic fish saved 6.62% of the total energy from eaten food for growth improvement. That phenomenon was named as "fast-growing and less-eating" effect. Growth and feed utilization by MThGH-transgenic F4 fish feeding with diets containing different protein levels had also been carried out. Protein and energy intakes were significantly higher in the transgenics than in the controls fed with 20% protein diet, and recovered energy, as a proportion of protein intake, was also significantly higher in the transgenics than in the controls fed with 40% protein diet.

It was thus concluded that at a lower dietary protein level, transgenics achieved higher growth rates mainly by increasing food intake; but at a higher dietary protein level, transgenics achieved higher growth rates mainly through higher energy conversion efficiency. That is to say, transgenics are more efficient in utilizing dietary protein than the controls, which leads to transgenics getting a significantly higher specific growth rate than the controls. For the body composition, it was revealed that the transgenic fish had body contents of dry matter 1.6%, and protein 2.2% to 4.3% higher than those of the controls, but contents of lipid 3.9% to 13.1% lower than that of the controls48. The apparent digestibility of amino acids tends to be higher in the transgenics than in the controls, especially in fish fed diets with lower protein levels. While taking a look at the proportion of amino acids in transgenics and controls, there was no difference whatever betwen 17 amino acids. Thus, transgenic fishes would have much more nutritious value than control fishes. It is reasonable that fishes with "fast growing and less eating" as well as "high protein content and low lipid content" traits will fulfil human increasing requirement on protein source from fishes.

Nevertheless, both the mouse metallothionein-1 (MT-1) gene promoter and the hGH structural gene are not suitable for the purpose of producing farming species of transgenic fish. It was urged to construct "all-fish" gene for transfer. Researchers have cloned both common carp-actin gene (CA)49, and grass carp growth hormone gene (gcGH)50, and subsenquetly made a new construct of pCAgcGH, an "all-fish" genomic construct with a powerful promoter of b-actin gene from common carp and the whole transcription unit of GH gene from grass carp.

In the spring of 1997, this construct has been microinjected into the fertilized eggs of Yellow River Carp (Cyprinus Carpio L.), and a batch of CAgcGH-transgenic was produced. As these fishes grew up to 5-month-old, body weight of the heaviest transgenic individual was 2.75 kg, while that for the controls was 1.1 kg. About 10% of the transgenics were over 2.0 kg, while the controls were about 0.7 kg on average. It is more exciting that the heaviest body weight of 17-month-old transgenics, 7.65 kg, was two fold and more than that of the control siblings [unpublished data]. Some experiments revealed further that the gene constructs with genetic elements derived from fish expressed more efficiently in fish cells than that from mammalian sp., e.g. the transcriptional activity of mouse MT promoter was only 1/2 of that of carp MT promoter in CAT (chloramphenicol acetyl-transferase) gene-transferred fish cells. Till now, more than ten "all-fish" recombinant genes have been constructed all over the world. The most dramatic growth acceleration came from Delvin et al., in which their pOnMTGH1-ransgenic salmon were more than 11-fold heavier than the controls on average.

It is very sensitive to talk about the biosafety of transgenic fish. There are food, genetic and ecological safeties, each of which should be concerned seriously. At present, the widely accepted principle on safety evaluation of foods produced by modern biotechnology was the "substantial equivalence principle" delivered by European OECD (Organization for Economic Cooperation and Development) in 1993. According to this principle, "all-fish" transgenic carp is included in "level I", the safest level. Studies from Cui's investigation revealed that the transgene could only flow among individuals within a species but not between species by natural reproduction. Fish crossbreeding has been widely used in aquaculture, two sets of whole genome of different species mixing with each other, i.e. 105 genes of one species crossing with 105 genes of another species. On the other hand, the "all-fish" gene transfer can also be regarded as a crossbreeding, but 105 genes from one species cross with 1 gene of another species instead.

It can be simply figured out that the hetrozygosity of genome in the CAgcGH-transgenic common carp is about 105 times less than that in the hybrids between common carp and grass carp. In other words, the risk of stocking "all-fish" gene transgenics is considerably on a lower level in comparison with stocking the hybrid fish. Nevertheless, people should take a cautious attitude towards the application of transgenics, since transgenic fish is still a "newborn" in comparison with naturally existing species that have undergone a long evolutionary and selective course. As polyploid-breeding in fish has been very popular during the past, this technique can also be employed in the breeding of transgenic fish. By crossing tetraploid individual with haploid transgenics, the infertile triploid strain of transgenic fish could be generated. It is reasonable to consider that stocking infertile strain of transgenic fish will lessen their impact on water ecosystem to the least degree.

THE PROSPECT OF TRANSGENIC FISH

Just as what has been discussed at the preceding part of this article, the transgenic fish generated at present is far from a genetic homogenous strain. Gene transfer in fish has not succeeded in the site-specific integration, controllable expression and stable transmission of the transgene. One of the most efficient ways to solve this problem is to use gene targeting technique in fish gene transfer. Gene targeting, homologous recombination between DNA sequence residing in the chromosome and newly introduced cloned DNA sequence, allows the transfer of any modified gene into the host genome of living cells. Since embryonic cell culture4 and embryonic cell nuclear transplantation techniques have been developed in fishes, gene targeting is hopeful to be carried out to gain embryonic cell lines carrying artificially modified and site-specific integrated gene. By nuclear transplantation with gene-targeted embryonic cells, the genetic homogenous strain of transgenic fish can be generated. Additionally, to establish stem cell like lines in some model fish species, such as zebrafish and medakafish, were successfully reported. If researchers can make great progress on the study of stem cell in farming fishes, it will be more convenient to generate homogenous strain of transgenic fish for aquaculture.

5

Industrial Works in Fish Industry

Consumption of seafood has increased over the last decade, without a concomitant increase in reported illness. This increased consumption trend is expected to continue both for prepared and for fresh or frozen varieties. The 1989 consumption figure was 15.9 pounds of edible meat per person per year. Total commercial landings were a record 8.5 billion pounds in 1989, and imported edible products totaled 3.2 billion pounds. The majority of the seafood supply was harvested from wild populations. The aquaculture portion of this supply will probably increase. A substantial amount of seafood (600 million pounds of finfish and 300 million pounds of shellfish) is caught recreationally. About 70% of commercially produced seafood in the United States is sold fresh or frozen. Canned seafood constitutes approximately 25%, and smoked/cured products 5%, of the seafood consumed.

The United States exported 1.4 billion pounds of edible domestic fishery products in 1989. The largest importer was Japan; Canada, the United Kingdom, France, and South Korea also provided good markets. The seafood harvest by industry is fragmented, diversified, seasonal, complex, and difficult to manage. Studies are needed to monitor changing :onsumption trends and patterns.

The processing, distribution and merchandising of finfish and shellfish will require more emphasis to reduce cross-contamination. Attention must be given to aquaculture in order to produce high-quality, consistently available species. Attention must also be focused on the harvesting, handling, distribution, and preparation of recreationally harvested fish to ensure consumer safety. More emphasis should be placed on educating the industry and the consumer about safe handling practices that can reduce potential food-handling problems.

INTRODUCTION

As Americans become increasingly aware of the relation between diet and good health, the consumption of fishery products will most likely increase.

The consumer recognizes that fish and shellfish are nutritious and wholesome foods. They are perceived as an excellent source of high-quality protein, containing lipids with high levels of unsaturated fatty acids, and perhaps contributing to the enhancement of human health by reducing the risk of cardiovascular disease. Likewise, seafood is characteristically tender, easily digested, and a good source of many important minerals and vitamins (NRC, 1989).

Although the attributes of seafood attract a more health conscious consumer, they also enforce expectations for enhanced safety. Consumer consciousness of seafood safety issues has become, as a result, increasingly important. Pollution and environmental issues have further focused people's attention on contamination problems. Concurrent media coverage and public interest groups have heightened the demand for rigorous safety standards in the food industry (Haas et al., 1986; Newton, 1989).

Unlike meat and poultry, which are derived from domesticated sources, the majority of the edible seafood supply in the United States, approximately 12.0 billion pounds including domestic landings and imports in round weight equivalents, was harvested from wild populations in 1989 (NMFS, 1990). The aquacultured portion of this supply is predicted to increase from both foreign and domestic sources (Redmayne, 1989), and recorded commercial landings are further supplemented by a growing portion of recreationally caught seafood destined for consumption.

Because the supplies of many seafoods are relatively small and regional, large numbers of individuals, using a variety of vessels that range from small boats to large factory ships, are involved. The seafood harvesting industry is highly fragmented. The diversity of the industry, the seasonal nature of fishing, the complexity of fish processing operations, and the substantial amount of seafood caught recreationally (600 million pounds of finfish and 300 million pounds of shellfish) make it difficult to manage and regulate these living resources (NOAA, 1990).

Both finfish and shellfish are subjected to contamination and cross-contamination in their natural habitat, as well as at any point during handling, processing, distribution, or preparation (Haas et al., 1986; Newton, 1989; NOAA, 1990). Seafood-borne illness has been reported due to natural toxins, microbial contamination, parasites, poor seafood handling, and chemical contaminants (CDC, 1981a-c, 1983a,b, 1984, 1985, 1989; FDA, 1989).

Because of the primary reliance on limited data-reporting systems via state departments of public health, and eventually the Centers for Disease Control (CDC), the extent of the public health risk due to cumulative exposure to microorganisms, natural toxins, and chemical contaminants cannot be assessed easily, especially in the context of total dietary exposure. Given this qualification, current data indicate a decrease in the reported incidents of illness from seafood relative to consumption. The committee has critically examined and evaluated the degree of severity of illnesses, their significance, and the extent of possible health risks involved. Its findings are documented in subsequent chapters of this report.

DEMOGRAPHICS OF THE SEAFOOD INDUSTRY

In 1989, commercial and recreational fishermen harvested more than 8.5 billion pounds of fish and shellfish from U.S. waters, which includes edible and industrial products. More than 300 major species of seafood were marketed, reflecting the diversity of the resource base (NMFS, 1990). Over 4,000 processing and distribution plants handled the commercial products of the nation's 256,000 fishermen. Almost 95,000 boats and vessels constituted the fleet (NMFS, 1990). Although commercial establishments are easily documented, the number of recreational fishermen and their support base are more difficult to quantify. Increasing numbers of anglers for fish from the nation's freshwater, estuarine, and marine waters are producing a growing share of the fresh and frozen seafood in today's diet. The number of recreational harvesters has been estimated to be in excess of 17 million individuals (NOAA, 1990).

Fresh and frozen seafood constitute about 70% of the product consumed in the United States. Canned seafood, particularly tuna, constitutes almost 25% of domestic consumption, and cured/smoked products account for the remaining 5% of per capita consumption.

FISHERY RESOURCES

Commercial landings (edible and industrial) by U.S. fishermen at ports in all the fishing states were a record 8.5 billion pounds (3.8 million metric tons) valued at $3.2 billion in 1989 (NMFS, 1990). This was an increase of 1.3 billion pounds (576,300 metric tons) in quantity, but a decrease of $281.8 million in value, compared with 1988. The total import value of edible fishery products was $5.5 billion in 1989, based on a record quantity of 3.2 billion pounds. Imports of nonedible (industrial) products set a record in 1989, with products valued at $4.1 billion, an increase of $676.1 million compared with 1988 (NMFS, 1990).

The trade deficit in fishery products has not declined. The dollar value of imports was higher in 1989 than in the previous year (NMFS, 1990). Canada is still the largest exporter to the United States, sending in more than 700 million pounds of fishery products in 1988. Ecuador was ranked second and Mexico third. Whereas Canada ships finfish products, shrimp is the primary commodity exported by Ecuador and Mexico. Imports from Thailand and China are both increasing due to rising shrimp production from their expanding aquaculture systems.

On a worldwide basis, aquaculture is becoming a major new factor in seafood production. The cultivation of high-value species, popular in the U.S. market, is a major factor in import sourcing. China, for example, along with other Asian nations, is replacing South and Central American countries as a major shrimp supplier to the United States. Aquaculture is expected to determine much of the future fisheries growth, because wild stocks are nearing full utilization (NMFS, 1990; NOAA, 1990).

The total export value of edible and nonedible fishery products of domestic origin was a record $4.7 billion in 1989, an increase of $2.4 billion compared with 1988. The United States exported 1.4 billion pounds of edible products valued at $2.3 billion, compared with 1.1 billion pounds at $2.2 billion exported in 1988. Exports of nonedible products were valued at $2.4 billion. Japan continues to be America's best export customer. Over 700 million pounds of seafood was sold to the Japanese market, with salmon, crabs, and herring the primary commodities. Canada, the United Kingdom, France, and South Korea were also good markets in 1989, but the value of their imports was small, compared to Japan's purchase of West Coast products (NMFS, 1990).

Consumers in the United States spent an estimated $28.3 billion for fishery goods in 1989, a 5% increase from 1988 (NMFS, 1990). The total included $19.1 billion in expenditures in food service establishments (e.g., restaurants, carryouts, caterers); $9.0 billion in retail stores (for home consumption); and $181.7 million for industrial fish products. In producing and marketing a variety of fishery products for domestic and foreign markets, the commercial fishing industry contributed $17.2 billion in value-added dollars to the gross national product (GNP), an increase of 5% compared to 1988.

Consumption of fish and shellfish in the United States totaled 15.9 pounds of edible meat per person in 1989 (NMFS, 1990). This total was up 0.7 pound from the 15.2 pounds consumed per capita in 1988. Per capita consumption of fresh and frozen products registered a total of 10.5 pounds, an increase of 0.3 pound from the 1988 level. Fresh and frozen finfish consumption was 7.1 pounds per capita in 1989. Fresh and frozen shellfish consumption amounted to 3.4 pounds per capita, with canned fishery products at 5.1 pounds per capita, up 0.4

pound over 1988. The per capita use of all fishery products (edible and nonedible) was 62.2 pounds (round weight), up 2.8 pounds compared with 1988 (NMFS, 1990).

Although most of the fish and shellfish consumed is from commercial production, a significant share is caught recreationally. In 1990, the National Marine Fisheries Service (NMFS) estimated that 17 million marine anglers harvested more than 600 million pounds of finfish (NOAA, 1990). Although statistics are lacking, NMFS suggests that 200-300 million live pounds of molluscs and crustaceans was harvested by recreationalists. This catch represents 3-4 live pounds or about 1-1.5 edible pounds of domestic per capita consumption (Krebs-Smith, 1989), outside the commercial figure of over 15 edible pounds per person. The source, handling, and distribution of the recreational catch are just beginning to draw attention. Indeed, because recreational anglers are not regulated as food producers/manufacturers, there is concern about the use and distribution of this "recreational" resource.

Although it is difficult to give definite numbers for either the commercial or the recreational harvesting sector, some general observations can be made. Commercially, the trend is toward more efficient activity. Consequently, the number of participants in the commercial sector is decreasing. The commercial processing industry appears headed toward consolidation, with increased dependence on imported products and aquaculture. Recreational participation remains strong. Consumption data, as suggested by both the Department of Agriculture and the Department of Commerce, indicate a continued, if not expanding, harvest of sport caught fish and shellfish. More than 20% of all fresh and frozen seafood consumed in the United States may now be attributed to noncommercial harvest and distribution.

AQUACULTURE

Aquaculture is a rapidly growing mode of production in the seafood industry. Annual production of farmed fish and shellfish in the United States has grown 305% since 1980 (TFTC, 1988). The greatest production is of catfish (Sperber,

1989). Catfish production increased 31% from 1986 to 1987. According to the Catfish Institute, farm-raised catfish increased from 5.7 million pounds in 1970 to 295 million pounds in 1988 and were expected to exceed 310 million pounds in 1989 (Sperber, 1989). Salmon production in the Pacific Northwest and Maine totaled 85 million pounds in 1987. In addition, other fish that are farmed include trout, redfish, sturgeon, hybrid striped bass, carp, and tilapia, as well as shellfish and crustaceans such as oysters and crawfish. Crawfish production acreage has increased 145% to about 160,000 acres. Overall U.S. aquaculture production of fish and shellfish increased from 203 million pounds in 1980 to some 750 million pounds in 1987. It is estimated that by the year 2000, that figure will reach 1.26 billion pounds.

Large amounts of cultured fish and shellfish are also imported annually. Approximately one-half of the 500 million pounds of shrimp imported is cultured (Schnick, 1990); 143 million pounds comes from China and Ecuador, neither of which regulates the use of chemotherapeutic agents in culture. More than 40 million pounds of salmon is also imported annually, often from countries similarly lacking tolerance levels for residues. Of special interest are the use of chloramphenicol in shrimp culture and ampicillin in yellowtail culture (Hawke et al., 1987; Manci, 1990). The Food and Drug Administration (FDA) has not examined imported seafood for drug residues, and there is no information regarding levels that might be ingested (Schnick, 1990). Aquaculture also produces fish used to stock recreational fishing areas. This procedure is under the control of government agencies that follow FDA regulations, use only approved drugs, and abide by legal withdrawal times.

CONSUMPTION TRENDS

Today's consumer is changing rapidly. Instead of single-income households, it is increasingly more common to have both man and woman working. The size of the family is decreasing. As many as one-fourth of all households are occupied by one person. This means more shoppers and diners, most with little time for home preparation (Davis, 1989).

Most adult men and women now work outside the home. In recent surveys, 7 out of 10 new home buyers noted that they will need two incomes to pay their respective mortgages. Nevertheless, the growth in two-income couples has generally created an increase in disposable income, but with little time to spend it. With as many as 50% of new mothers working outside the home within the first year of childbirth, it is easy to see the revolutionary changes taking place among families. The working mother or single dweller does not have the time to prepare meals in the traditional sense. In recent Food Marketing Institute (FMI) surveys, more than 30% of the husbands of women who work full-time did as much cooking, cleaning, and food shopping as their wives (Davis, 1989; FMI, 1988). The population is aging. Going into the next century, the fastest growing groups will be those aged 45 to 54, along with those over age 85. By the year 2000, the proportion of Americans over age 65 will be the same throughout the country as the proportion in Florida today. An aging population means decreased discretionary spending and more demands for healthful and nutritious foods.

Minorities are growing in America. Within 10 years, one-quarter of all Americans will be either black, Hispanic, or Asian. The city of Los Angeles illustrates the trend. At present, Los Angeles is the largest Mexican city outside Mexico, the second largest Chinese city outside China, the second largest Japanese city outside Japan, and the largest Philippine city outside the Philippines (Davis, 1989).

The consumer demand for convenience, gourmet foods, ethnic items, and other services is increasingly evident in the food service and retail food industries. As the number of working women and single dwellers increases, the consumer base continues to change. With reduced leisure time, consumers who once spent two hours per day in the kitchen now spend less than a half hour. Convenience stores, fast-food restaurants, specialty food service outlets, and prepared items in the supermarket are food industry responses (FMI, 1988; Taylor, 1989). To illustrate the impact of less preparation time in the home, a quick review of consumer buying habits is in order.

In 1973, almost 80% of the food dollar was spent on home-prepared foods. In 1988, this number had fallen to 67%. Many predict that the figure may be as low as 40% by the year 2000. As with all foods, fish and shellfish preparation must be viewed in the manner in which consumers use the product in a contemporary environment. This does not mean that the consumer will be eating at home less but, rather, that less time will be devoted to food preparation. This trend toward "cocooning," in which the family spends more time around the home but utilizes the time more prudently, is central to future consumer patterns (Davis, 1989).

Consumers want more convenience and nutrition. Value-added products, ready-to-eat items, and microwave entrees are examples. Deli departments of the supermarket may soon become food service operations, competing with fast-food and takeout restaurants (FMI, 1989; Taylor, 1989).

Seafood, like other foods, will be placed in a competitive consumer environment. Fish and shellfish must continue to taste good if they are expected to attract more consumers. Further, seafood must stay within the budget of the new consumer. If the industry can respond to the changing consumer base, the opportunity to expand per capita consumption appears good (Taylor, 1989). The amount of imported product is not yet recognized as a potential problem by the consumer, yet it is of significant concern to regulatory officials. Rising needs place increased pressure on government to protect consumers without the ability to monitor the harvest, processing, and distribution of the hundreds of species in question. Because of the potential of ever-increasing imports, the safety issue is becoming a matter of international concern. Although agencies routinely sample and require country-of-origin labeling, the consumer is unaware of the complexity of attempting to truly safeguard these foodstuffs.

ACTIVITIES IN OTHER COUNTRIES

A number of countries have endeavored to enhance the value of their seafood products by enacting programs to ensure product quality. Canada, Denmark, and Norway have given high priority to marketing safe, quality seafood items. Canada, for example, inspects vessels, landing sites, and processing facilities on an annual basis. Vessels must meet the same exacting standards as processing facilities or risk losing their certification. Canadian plant registration requires compliance with a posted list of standards. At inspection, plants are rated by use of a Hazard Analysis Critical Control Point (HAACP) approach. Critical findings result in more frequent inspections or the possibility of noncertification.

In Europe, similar programs are in place. Denmark inspects fishing vessels. Each participant must meet certain sanitation requirements, as well as certification for activities such as on-board processing. Distribution centers receive regular inspections that monitor all products entering the marketplace. The advent of the European Economic Community (EEC) has brought forth a host of new regulations, ensuring that member nations comply with the policies of their EEC partners.

Many other countries have seafood inspection programs, but they are often not dedicated programs like those in Canada, Denmark, Norway, Iceland, and New Zealand. Consequently, they do not pay the same rigorous attention to detail. Indeed, most countries have programs centered on seafood as a food group, not as a distinct entity that requires special attention.

CONCLUSIONS AND RECOMMENDATIONS

Based on commercial sources, Americans consumed almost 23% more seafood in 1989 than they did 10 years earlier. This increase in consumption was not accompanied by a concomitant increase in reported seafood-borne illnesses. The total supply of fishery products to fulfill the domestic requirement for seafood was in excess of 8.5 billion pounds in 1989, with over 300 species involved in the catch statistics. Production and consumption trends suggest that domestic seafood demand will continue, with more emphasis on prepared convenience foods along with the traditional demand for fresh and frozen selections. Production will have to be supplemented with more imported and cultured sources. Recreational harvesting, both in the purist sense and as subsistence fishing, continues to contribute a significant portion to the annual per capita intake. The committee recommends the following:

- Consumer information studies must be conducted to monitor the rapidly changing consumption trends in the United States. Patterns of consumer use and preparation, as well as sources of seafood products used in the home, must be evaluated. By better understanding consumption patterns, fishery managers and food regulators will be more able to influence dietary intake, and reduce potential exposure to fish from contaminated water.
- Changes in consumption patterns necessitate more attention to informing consumers on how to best handle highly perishable products such as seafood. As much as 50% of all reported, acute fish and shellfish problems might be eliminated by more careful handling and proper preparation in the home or in food service establishments. With the advent of more prepared foods, every effort should be made to ensure the safety of the product both in the manufacturing/distribution chain and for the end user.
- The retail and institutional handling of seafood products requires increased attention to control cross-contamination. A number of seafood-related illnesses can be traced to poor sanitation practices by employees or to lack of proper handling via the distribution system. More efforts will be needed to alert all users to the importance of time/temperature relationships, HACCP concepts, good manufacturing practices, and new technology (e.g., live holding tanks).
- Aquaculture promises to produce a larger share of domestically consumed fish and shellfish in the years ahead. Cultured plants and animals hold the promise of being high quality, and generally free of some of the contamination associated with wild species. Care, however, must be taken to avoid the untimely use of antibiotics and other chemicals in these closed or recirculated systems, which are often used to control pathogens in semiclosed systems.

- The safety of recreationally harvested fish and shellfish requires increased vigilance, which means increased focus on the origin, handling, and distribution of recreational products. These harvesting efforts may now account for over 20% of all fresh and frozen seafood consumed in the United States. However, this catch is not well controlled, and users may handle, distribute, and prepare the product in an unsafe manner. Further, much of this product may be harvested from areas not suited for consumption due to natural or induced contamination problems. Increased educational activity is required to protect the consumer with regard to this resource. Fishery managers will have to pay greater attention to the implications of sport caught fish and shellfish on consumer health.

EXISTING PRACTICES AND SUGGESTIONS FOR IMPROVEMENT

The difficult issue of determining human intakes for a broad (though far from comprehensive) range of chemical contaminants in seafood is subsequently addressed. Estimates are made of national average daily intakes of various inorganic and organic contaminants via commercially marketed seafood, and for several organic carcinogens, upper-confidence-limit estimates of possible cancer risk are made. However, aside from the methylmercury example,[1] in the absence of better information on the population distribution of the dosage of contaminants to the U.S. population, it is impossible to make even tentative quantitative estimates of potentially significant noncancer risks. Of additional serious concern are the appreciable quantities of seafood consumed following noncommercial sport and subsistence tribal fishing.

Finally, an overview of opportunities for research on different categories of potential health impacts is presented, and conclusions are drawn from both this and the previous chapter. The principal conclusions are the following: From both natural and human sources, a small proportion of seafood is contaminated with appreciable concentrations of potentially hazardous organic and inorganic chemicals. Some of the risks that may be significant include reproductive effects from PCBs and methylmercury, and carcinogenesis from selected PCB congeners, dioxins, and some chlorinated hydrocarbon pesticides.

Consumption of some types of contaminated seafood poses enough risk that efforts toward evaluation, education, and control of that risk must be improved.

Present quantitative risk assessment procedures used by government agencies can and should be improved and extended to noncancer effects.

Current monitoring and surveillance programs provide an inadequate representation of the presence of contaminants in edible portions of domestic and imported seafood, resulting in serious difficulties in assessing both risks and specific opportunities for control.

Because of the unevenness of contamination among species and geographic areas, it is feasible to narrowly target control efforts and still achieve meaningful reductions in exposures.

The data base for evaluating the safety of certain chemicals that find their way into seafood via aquaculture and processing is too weak to support a conclusion that these products are being effectively controlled.

The principal recommendations of the committee are as follows:

Existing regulations to minimize chemical and biological contamination of the aquatic environment should be strengthened and enforced.

Existing FDA and state regulations should be strengthened and enforced to reduce the human consumption of aquatic organisms with relatively high contaminant levels (e.g., certain species from the Great Lakes with high levels of PCBs, swordfish and other species with high methylmercury levels).

Federal agencies should actively support further research to determine the actual risks from the consumption of contaminants associated with seafood and to develop specific approaches for decreasing these risks.

Increased environmental monitoring should be initiated at the state level, as part of an overall federal exposure management system. States should continue to be responsible for site closures, and for issuing health and contamination advisories tailored to the specific consumption habits, reproductive or other special risks, and information sources of specific groups of consumers. There should be an expanded program of public education on specific chemical contaminant hazards via governmental agencies and the health professions.

Part of the committee's charge was to review and summarize "the current status of regulations, guidelines, and advisory statements issued by Federal and State public health authorities on environmental contaminants in seafood." Its review was to specifically address contaminants defined by Food and Drug Administration (FDA) regulations as "avoidable or unavoidable." Then, based on this, the committee was asked to "assess how well the current regulatory framework protects the public health." The committee was also charged with the task of reviewing and summarizing, specifically, the health risk assessment procedures used by FDA, the Environmental Protection Agency (EPA), and other regulatory authorities for priority environmental pollutants, including toxic metals and synthetic organic chemicals. In addition, the committee was asked to "recommend future research directions, as appropriate."

To set the stage for an examination of how current risk assessment procedures can be improved, the basic concepts underlying the mechanisms of action of toxic substances are articulated in the following section, along with quantitative ideas about dose-time-response relationships.

Then an extensive critique of agency risk assessments for PCBs and methylmercury is provided. Finally, the committee addresses issues of exposure assessment and risks from other substances, and opportunities for further research on potential chemical health hazards.

BROAD CATEGORIZATION OF MECHANISMS OF DIFFERENT ADVERSE EFFECTS AND IMPLICATIONS FOR DOSE-RESPONSE RELATIONSHIPS

The concern from which much of our regulatory history has resulted is the potential carcinogenic effect of some contaminants. To a certain extent, this concern is based on the mutagenic mechanisms of cancer. For carcinogens that act by primary genetic mechanisms, there are good theoretical reasons to believe that at the limit of low dosage, the risk will be a linear function of exposure (Ehrenberg et al., 1983; Hattis, 1990a). However, as more is learned about the mechanisms of some other types of toxic effects–particularly reproductive effects and chronic degenerative neurological conditions (NRC, 1989, 1990)–concern about potential low-dose effects of other types has tended to increase. It is therefore important to clarify what the expectations should be for dose-response relationships from first principles, given the full range of causal processes that can lead to impairment of health.

Are the events occurring ordinarily fully reversible (or very nearly so), given a prolonged period with no further exposure to the hazard?

Types of Health Hazards Requiring Fundamentally Different Risk Assessment Approaches

1. "Traditional" toxicity resulting from overwhelming body compensatory processes: below some threshold, in individuals who are not already beyond the limits of normal function without exposure, response is reversible.

- Traditional acute toxicity–Toxic action is completely reversible or proceeds to long-term damage within about three days of exposure (paralytic shellfish poisoning, puffer fish poisoning; probably many teratogenic effects).
- Traditional chronic toxicity—Toxic process typically proceeds to permanent damage over a period of several days to several months, due to either (1) reversible accumulation of a toxic agent (e.g., methylmercury, lead) or (2) accumulation of a slowly reversible toxic response (e.g., cholinesterase inhibition).

2. Effects resulting from insidious processes that are irreversible or poorly reversible at low doses or early stages of causation.

- Molecular biological (stochastic process) effects—Effects occur as a result of one or a small number of irreversible changes in information coded in DNA: mutagenesis, most carcinogenesis, and some teratogenesis.

- Chronic cumulative effects–Effects occur as a result of a chronic accumulation of many small-scale damage events: emphysema, noise-induced hearing loss, atherosclerosis, and probably hypertension; possibly depletion of mature oocytes.

If the answer to the first question is yes, then it will generally be appropriate to treat the condition within the framework of traditional toxicology.[3] Some examples of such reversible changes are the following: Buildup of a contaminant in blood or other tissues. It is rare for there to be a zero rate of excretion of any material. Given time and no further exposure, toxicant buildup should be reversible, although it can be quite prolonged. [Current estimates are that only about 9% of more persistent polychlorinated biphenyl (PCB) isomers may be metabolized or excreted per year in humans (Yakushiji et al., 1984).]

Most enzyme inhibition (generally, even irreversible inactivation of enzyme molecules can be "reversed" through the synthesis of replacement molecules)

Induction of short-term biological responses that act to maintain homeostasis (e.g., sweating in response to heat, tearing in response to eye irritation)

If the answer to the above question is no and events are likely to be occurring at subclinical exposure levels or preclinical stages that are not ordinarily reversible, the modeling of biological risks will have to be based on concepts that are fundamentally different from the homeostatic system/threshold paradigm. Examples of such irreversible or poorly reversible events include:

- changes in genetic information or the heritable pattern of gene expression after these are effectively "fixed" into a cell's genome expression by replication;
- death of nonreplicating types of cells (adult neurons);
- destruction of nonregenerating structures (alveolar septa); and

- generation and buildup of incompletely repaired lesions (atherosclerotic plaques).

Appropriate modeling for conditions that are the result of irreversible or poorly reversible processes must be based fundamentally on the likely dose-response characteristics of the events that cause the basic irreversible changes.

Once the primacy of such changes is established for a particular event, the analyst should then ask whether clinical manifestations are likely to be the direct result of only a few, or very many, individual irreversible damage events. If only a few events are believed to contribute directly to a particular clinical manifestation (e.g., a small number of heritable changes within a singe cell line leading to cancer), the effect can be considered a "molecular biological" disease.

The risk assessment models used must follow from an understanding of the stochastic nature of the basic process. On the other hand, if thousands, millions, or billions of individual irreversible events directly contribute to a particular condition (e.g., very large numbers of individual neurons must die to cause the clinical manifestations of Alzheimer's or Parkinson's disease), the biological harm should be dealt with under the novel category of chronic cumulative conditions.

Traditional Acute Toxicity

Three kinds of insights for acute toxicity risk assessment follow naturally from the homeostatic system paradigm of physiology and traditional toxicology:

1. There will be a series of toxic effects as different compensatory processes are overwhelmed and as impairment broadens from more-to less-sensitive cells and functions.
2. For each effect in each individual who is not already beyond the limits of normal functioning in the absence of exposure, there will be some subthreshold level of exposure that will be insufficient to produce the effect.
3. Individuals will differ in their thresholds.

A caveat to the general expectation of individual thresholds is that some tasks may so tax the capabilities of a system (perhaps, during fetal life, the struggle to mobilize metabolic resources to grow and differentiate as fast as possible so as to cope with the external world at birth) that any impairment of a key limiting functional parameter required for the task could compromise function to some degree. (This would also apply to reaction time for a driving task, for example.) Of particular relevance to the committee's task in this regard is the suggestion of some studies that dietary PCB exposure may be associated with either changes in birth weight (Fein et al., 1984; Sunahara et al., 1987; Taylor et al., 1989) or indices of neurological function in infants (Jacobson et al., 1985; Rogan et al., 1986).

The first job in assessing acute toxic effects is to define the series of acute responses to the disturbing influence in question. Ideally, the analysis should then attempt to determine (to whatever degree of precision is possible) the nature and magnitude of the dosage and the disturbance of physiological parameter(s) that are necessary to cause each type of acute toxic response, along with the frequency of each response in a diverse human population.-Such mechanism-based analysis is, however, not common in the field. Rather, the current state of the art in those rare cases where acute toxic effects are treated quantitatively is to use probit equations (Finney, 1971) of the general form

Prohibit of response = a + b ln (Cn T),

where *a, b,* and *n* are constants, *C* is external concentration, and *T* is exposure time; *n* represents the basic trade-off between intensity and duration of exposure, and *b* defines the breadth of an assumed lognormal distribution of threshold responses. Although some of the available animal data on irritant gases appear to be well summarized by equations of this form (Appelman et al., 1982), this is basically an empirical formula that does not incorporate quantitative representations of the various processes underlying toxicity. It is therefore difficult to decide what adaptations should be made in applying the empirical relationships to diverse subsets of humans.

Even more common, unfortunately, is the simple use of the no-effect level (NOEL)/"safety factor" analysis for arriving at acceptable daily intake (ADI) levels for chemical contaminants.

Rather than estimate the numbers of people with specific degrees of particular effects, the general approach is to arrive at an ADI by a rule-of-thumb procedure derived from observed NOELs [or, more recently, no-adverse effect levels (NOAEL), after often contentious discussion over what effects are considered "adverse"], or sometimes low effect levels (LOELs) in animal experiments or human studies.

When projections are made from animal data, generally a 100-fold "uncertainty factor" is allowed between the NOEL and the ADI (Dourson and Starra, 1983). The 100-fold factor is often decomposed as 10-fold to account for possible differences in sensitivity between humans and the most sensitive species tested and 10-fold to account for possible interindividual differences in susceptibility among humans.

This approach has a few advantages:

1. It is "quick and dirty"—relatively straightforward to apply and does not require complicated model building or analysis.
2. Through thousands of applications in the past, it is not yet known to have led to catastrophic adverse effects in humans (using Ozonoff's working definition of a "catastrophe" as an effect so large that even an epidemiological study can detect it (D. Ozonoff, Boston University School of Public Health, personal communication, 1990).

On the other hand, for the long term, the simple uncertainty factor approach has a number of disadvantages:

1. No one knows how protective it really is, either in general or in specific cases. What fraction of the diverse human population can be expected to experience adverse effects when exposed at the level calculated to be "acceptable" under the formula? (In general, there may be some finite fraction of individuals who, because of disease or other reasons, are marginal for biological functions affected by the chemical and who may be pushed beyond a functional threshold for an adverse effect by a small finite dose of the chemical.)

2. The procedure incorporates one specific social policy standard for setting "acceptable" levels without making clear where technical analysis leaves off and policy/value analysis begins.
3. Effects are generally scored as either present (operationally, statistically significant) or not present (not statistically significant) at a particular dose. There is usually no quantitative analysis of the effects of sample size or the dose-response relationship for the effect in question.
4. There is no defined or obvious way to incorporate newer types of relevant data on human interindividual differences in rates of uptake/absorption for a constant environmental exposure ("exposure variability");

- rates of activating or detoxifying metabolism and excretion, producing differences in the concentration x time of active metabolites per unit of absorbed dose at the site of toxic action ("pharmacokinetic variability"); and
- differential risk of response ("response variability") for a given concentration x time of active metabolites at the site of toxic action.

In particular, the committee suspects that the inability of the uncertainty factor paradigm as usually formulated to incorporate newer types of relevant information into a systematic procedure for updating assessments of health hazards has tended to discourage both the collection and the analysis of potentially important data. One example of this is information on human interindividual variability in parameters that could affect susceptibility.

Such a parameter is generally a continuous variable that has a strong causal influence on performance of an important biological function (although it will not generally be the sole determinant of performance). It should, in turn, be affected by the toxin/exposure under study, and it should be reasonably likely that effects on the final health condition of concern are primarily mediated through effects on this functional intermediate parameter. For example, a key functional intermediate for some reproductive effects of PCBs may well be changes in birth weights (Fein et al., 1984; Jacobson et al., 1985; Sunahara et al., 1987; Taylor et al., 1989). Similarly, blood or tissue concentrations of lead constitute a useful intermediate parameter for lead toxicity.

Elements of a New Analysis for Noncancer Health Effects Mediated by a "Functional Intermediate" Parameter

1. Elucidate the quantitative relationships between internal dose/time of toxin exposure and changes in the functional intermediate parameter.
2. Assess the preexisting "background" distribution of the functional intermediate parameter in the human population.
3. Assess the relationship between the functional intermediate parameter and diminished physiological performance or adverse health effects.
4. Assess the magnitude of parameter changes likely to result from specific exposures in humans (taking into account human interindividual variability in metabolism and other determinants of pharmacokinetics) and consequent changes in the incidence and severity of health effects.
5. Do not attempt, from the biology alone, to determine "acceptable" levels of parameter change or exposure. (Let policymakers decide what changes in the incidence and severity of health effects are "acceptable" in the context of modes of exposure and in light of the feasibility of reducing or avoiding exposure.)

In principle, the use of such intermediate parameters can provide windows on the pathological processes that occur earlier in the development of toxicity, are more sensitive to the action of potential toxicants (compared with attempts to observe actual cases of illness), and are more accessible to direct comparative measurement in both animal models and humans. It is desirable, for these purposes, that the intermediate parameters chosen be as close as possible to the actual causal pathway leading to harm. However, even a parameter such as birth weight, which may not itself bear a direct causal relation to infant deaths, may be a close enough indicator of the actual causal processes to serve as a useful intermediate predictor.

Because there will generally be a series of steps in the sequence between toxin uptake and ultimate manifestation of adverse effects, the analyst may often have choices of which parameter(s) to use for assessing human risk. These choices will generally be based on the availability of measurement techniques and theory for observing or estimating the parameter(s) in question.

Traditional Chronic Toxicity

The basic principles that govern the analysis of acute toxic effects are by and large directly applicable to cases of chronic toxicity. Chronic toxic analyses tend to differ from analyses of acute toxic effects primarily in that considerable emphasis must be placed on the "slow step" of the process, which causes the effect to be chronic rather than acute. This slow step is generally either a long-term accumulation of a toxic agent that is poorly excreted under ordinary conditions (e.g., lead, mercury) or an accumulation of some slowly reversed residual effect (e.g., acrylamide). With lead as an example, the following statements can be made: Appreciable information is available on the pharmacokinetics of lead absorption, transport, storage, and excretion (Barry, 1975; Bernard, 1977; Campbell et al., 1984; Chamberlain, 1985; Marcus, 1985a-c; Rabinowitz et al., 1976); and much better information could be obtained with the aid of natural experiments such as strikes among lead-exposed workers, which can given information on the rate of decrease of blood lead levels after a reduction in exposure (Hattis, 1981).

Inhibition of heme synthesis enzymes at essentially all dose levels is well characterized (Haeger-Aronsen et al., 1974), and the inhibition of heme synthesis may be important in producing some of the neurotoxic effects of lead (Silbergeld et al., 1982), although the short- and long-term functional significance of different degrees of inhibition in different individuals is far from clear.

Effects on some measures of neurological function and kidney function are susceptible to study in reasonably straightforward ways. Effects on higher-order development of central nervous system functions are more difficult to determine because of an ignorance of basic mechanisms; however, some good studies have become available in recent years (Baker et al., 1983; Bellinger et al., 1987, 1990; HHS, 1988; Needleman et al., 1979, 1990; Waternaux et al., 1989). The impairment of very complex neurological functions by lead raises a significant issue in the application of the traditional toxicological paradigm to risk analysis. As indicated above, the usual assumption is that there is some functional reserve capacity in "normal" individuals that maintains "adequate" performance despite a "small" degree of perturbation of a biological parameter by a "low" dose of toxic material. However, if the function is already taxed to its limit in certain situations, even in the absence of exposure (perhaps for a first grader learning to read or for a developing fetus mobilizing all its available metabolic energy to grow and differentiate), and if the biological parameter being perturbed is limiting to the performance of that function, then any level of exposure may produce at least some reduction in performance. Addressing the issues of the population distribution of different functional reserve capacities, and the relationship of functional reserve capacities to specific biochemical parameters, is essential to the future research needs of risk assessment for classical chronic toxic agents. Also in the area of neurotoxicology, Silbergeld (1982) has written of the potential of new radioimmunoassay and functional measurement techniques to help shift the focus of research away from traditional morphological criteria of neurological damage toward more sensitive and sophisticated measures of performance.

Molecular Biological (Stochastic Process) Diseases

In addition to most carcinogenesis, molecular biological diseases include mutagenesis and at least some teratogenesis. The subject of quantitative risk assessment for carcinogenic hazards has been discussed extensively elsewhere (American Industrial Health Council, 1987; Bishop, 1987; Crump et al.,

1976, 1977; EPA, 1986a; Hattis, 1982; Hattis and Smith, 1986; Moolgavkar, 1986; Rai and Ryzin, 1981; Whittemore, 1980). However, it is worth briefly recapitulating some basic features of the carcinogenic process and their implications for cancer dose-response relationships.

Science is now much closer than it was a decade ago to understanding the fundamental mechanisms involved in carcinogenic transformation. For some time it has been clear that tumors arise as a result of a series of changes or rearrangements of information coded in DNA within single cells (Cleaver and Bootsma, 1975; Fialkow, 1977; Hattis, 1982; Knudson, 1973, 1977; McCann et al., 1975; Vogel and Motulsky, 1979). These changes are often induced by electrophilic metabolites of the parent compounds to which organisms are exposed (Miller and Miller, 1981). With the identification of "oncogenes," some detailed molecular characterization is being provided of the changes resulting in DNA (Fischinger and DeVita, 1984; Hoel, 1985; Modali and Yang, 1984; Yunis, 1983). It has also been apparent for some time that further headway cannot be made in elucidating the shapes of carcinogenesis dose-response relationships at low dosages simply by increasing the numbers of animals studied in conventional bioassays. A variety of mathematical models with dramatically different consequences for low-dose risk can always be found that fit the observations about equally well (Maugh, 1978; Whittemore, 1983). Low-dose risk projections are, therefore, inevitably much more determined by the choice of model than by the available data (Guess et al., 1977; Whittemore, 1980), if what is meant by "data" is restricted to observations of the incidence of ultimate adverse effects in small groups of animals.

Because of sample size limitations, animal carcinogenesis bioassays must be done within a limited range of relatively high dose levels. Typically, the difference between the minimum detectable response and a response that effectively saturates the system or causes interference through overt toxicity is only one to two orders of magnitude (often even less). Over this high dose range near levels where the agent produces overt toxic effects, enzyme saturation and other forms of pharmacokinetic nonlinearities are most likely. If in dose-response modeling for risk assessment, the nonlinearities of pharmacokinetic origin are not separated from the nonlinearities that may arise from the multiple mutation mechanism that is central to carcinogenesis, our ordinary curve-fitting procedures will implicitly attribute the pharmacokinetic nonlinearities to the fundamental carcinogenic process (Hoel et al., 1983). The resulting errors are particularly serious if one wishes to produce the best point estimates of carcinogenic risk in addition to upper confidence limits.

Clearly, to make real progress in modeling carcinogenic risks, knowledge of the fundamental processes involved must be used to break open the black box between external exposure levels and ultimate production of tumors. The use of pharmacokinetic models and intermediate parameters ("markers") to characterize the dose-response characteristics of small segments of the causal pathway to carcinogenesis has considerable potential to improve dose-response modeling for the process as a whole (Hattis, 1988). Such markers can include both those that may lie directly along the causal pathway, such as DNA adduct formation, and putative correlates, such as hemoglobin adduct formation, that can be good indicators of the concentration-time product of active intermediates in the systemic circulation.

One key fact must be recognized from the beginning about pharmacokinetic modeling, however. Whatever nonlinearities may be produced at high doses by the saturation of enzymes, the saturation of active transport processes, the depletion of cellular reactants for electrophilic agents, or changes in cell division rates to make up for cell killing due to overt toxicity, all of these nonlinearities must necessarily disappear as one approaches very low dose rates (Hattis, 1990a). The slope of the line relating ultimate DNA lesions in replicating cells to external dose may well be very different at low than at high doses, but it must be linear. The basic reason for this is that at low doses the rates of the transport and transformation processes that lead to DNA damage and repair depend directly on the number of collisions between molecules of an "input" chemical (or activated intermediate or DNA adduct) and a resident cellular reactant (or hole in a membrane or repair enzyme molecule). At low doses the number of resident cellular reactant molecules does not change appreciably as a function of the concentration of the input. Therefore, the number of relevant collisions and the rates of reactions and side reactions in the causal sequence at low dosage must be direct linear functions of the amounts of input chemical and its activated derivatives. Some finite fraction of the ultimate DNA lesions must escape repair before the next cell replication as long as the cells affected have a nonzero turnover rate, there are a finite number of repair enzyme molecules, and the repair molecules operate at a finite rate.

All carcinogens – in particular the PCBs and dioxins – are not thought to act primarily by causing DNA mutations (Safe, 1989). There is disagreement about how to estimate risks from carcinogens. In the United States, regulatory agencies generally estimate risks in the same way for both genotoxic and non-genotoxic carcinogens. Regulatory agencies adopt this position because they believe that full knowledge of the mechanism of action of non-genotoxic agents is needed before they can be assumed to exhibit thresholds. In several foreign countries, non-genotoxic carcinogens are generally assumed to have thresholds below which there is expected to be no risk. The disagreement is not confined to official government positions; some scientists prefer to treat non-genotoxic carcinogens as having thresholds, some do not.

INDIRECT MECHANISMS OF CARCINOGENESIS

Indirect processes that enhance the rate of "initiation" (the initial change or rearrangement of information in DNA that places a cell on a pathway to cancer):

1. Changes in basic transport processes (e.g., low-fiber diets may prolong the residence time of feces in the gut, leading to greater exposure of the intestinal epithelium to reactive agents).
2. Changes in metabolic processing (e.g., induction of enzymes such as some mixed function oxidases that

convert chemicals to forms that can directly react with DNA).

3. Changes in the effective amount of target tissue available for carcinogenesis (e.g., by simple hyperplasia).
4. Changes in the efficiency of DNA repair [e.g., inhibition of DNA repair by some metals (Zakour et al., 1981) or enhancement of cell replication (leaving less time for repair before a DNA lesion can be fixed into the genome as a permanent mutation)].

Indirect processes that alter the frequency with which "initiated" cells progress through subsequent stages in the carcinogenic process:

1. Induction of subsequent genetic changes: Many promoters appear to be capable of inducing the expression of Epstein-Barr virus antigens (Takada and Zur Hausen, 1984). Additionally, some "promoters" reportedly lead to the generation of active oxygen species that may damage DNA and lead to subsequent somatic mutations along the pathway to carcinogenesis without themselves being converted to compounds that react directly with DNA (Kinsella and Radman, 1978).
2. Changes in the frequency with which initiated cells are effectively removed by terminal differentiation [e.g., effects of early and multiple pregnancies in reducing later breast cancer risk (Kampert et al., 1988; Layde et al., 1989; Moolgavkar et al., 1980)].
3. Release of initiated cells from growth control by neighboring cells
   - Mimicry of the action of a growth regulator or hormone by an introduced substance. [Phorbol esters alter the binding and phosphorylation of epidermal growth factor receptors (Friedman et al., 1984; McCaffrey et al., 1984). The very high-affinity binding of 2,3,7,8-tetrachlorodibenzodioxin (TCDD) to the AH receptor induces an increase in transcription of the genes that code for certain P-450 isozymes, among other effects (Whitlock, 1989).

Genetic studies suggest that the AH receptor is necessary but not sufficient for the activity of halogenated aryl hydrocarbons as skin tumor promoters in hairless mice (Safe, 1989).]

- Inhibition of the action of normal growth suppressing substances. [Phorbol esters inhibit the binding of somatostatin (Zeggari et al., 1985).]
- Inhibition of the passage of growth inhibitors among cells by many chlorinated aromatic compounds (Tsushimoto et al., 1983).
- Killing of neighboring cells responsible for repression of initiated cells.
- Induction of cell replication among initiated cells, interfering with the ability of repressors to pass tight junctions or isolating some daughter cells from tight junctions.

4. Changes in the rates of proliferation or survival (without terminal differentiation) of initiated cells relative to the proliferation and survival of normal cells (theoretical mechanism suggested by Moolgavkar and Knudson, 1981).

Indirect processes that might alter the survival, growth and spread of tumors, or the progression of tumors to increased malignancy:

1. Changes in hormonally mediated processes that might speed up the growth of specific cell types (e.g., estrogens and some breast cancers).
2. Changes in the efficiency of immune surveillance in destroying incipient tumors at early stages. Some observations suggest that tumor promoters may alter the functioning of "natural killer" cells (Kabelitz, 1985). Immunosuppressive effects have also been observed for some promoters in vivo (Pasquinelli et al., 1985).
3. Changes in local tissue conditions that favor colonization of new tissues by metastases (e.g., the establishment of tumor blood supplies).

Generally, U.S. regulatory agencies adopt linear, no-threshold models and do not depart from them unless there is overwhelming scientific evidence to show that they are incorrect in specific cases. This is a prudent posture to take, until evidence of the details of specific nongenetic mechanisms is detailed enough for different low-dose assumptions to be indicated very clearly.

Chronic Cumulative Conditions

Chronic cumulative conditions include neurological conditions caused by the cumulative loss of neurons (e.g., Parkinsonism, Alzheimer's disease), emphysema and other chronic lung diseases, atherosclerosis, and hypertension. This new category is required because the underlying mechanism of these conditions – slow accumulation of many irreversible or poorly reversible damage steps – departs significantly from paradigmatic mechanisms that provide the basis for the three other, more traditional categories. These conditions are likely to be increasingly important as the population shifts to progressively older ages, and as continued progress is made in preventing and treating diseases that afflict younger people. For example, from 1969 to 1985, the percentage of nursing home residents with a primary diagnosis of a mental disorder rose from 11 to 22% (Hing, 1989, p. 9).

Theoretically, the approach that should be taken in developing risk assessment models for individual conditions within this new category is to:

1. describe the fundamental mechanism(s) that causes individual damage events to accumulate (especially the quantitative significance of various contributory factors);
2. elucidate quantitatively the ways in which specific environmental agents enhance the production or prevent the repair of individual damage events; and
3. describe the relationships between the numbers, types, and physical distribution of individual damage events and the loss of biological function or clinical illness.

Unfortunately, no examples are known in which there has been successful quantitative modeling of any of these three types for any chronic cumulative condition. Often, the qualitative nature of individual damage events is not difficult to discern. For example, atherosclerotic lesions are thought to be produced by a sequence of events described by Ross and Glomset (1976). However, no one has yet successfully developed a predictive model that relates the frequency or severity of these events to the various internal and external causal factors that must be involved. This is clearly an area that requires basic biomedical research, as well as creative interaction between mathematical modelers and experimentalists.

A couple of consequences of importance here, however, follow from the basic nature of the cumulative irreversible processes that define this category. First, it is important to recognize that the damage processes must be proceeding continually under quite usual everyday situations. This requires viewing with some suspicion the usual presumption of the traditional toxicological paradigm that the common adaptive responses to everyday insults are harmless. For diseases such as atherosclerosis, which proceed silently and chronically in ordinary individuals throughout their lives, homeostatic protective mechanisms must be failing in subtle ways rather frequently. It is likely that there are thresholds that give rise to the small-scale damage events of chronic cardiovascular disease processes [for example, perhaps the lining of the arterial wall in a particular region only suffers appreciable damage when systolic blood pressure temporarily goes above 180 millimeters (mm) of mercury (Hg)]. However, whatever thresholds exist must be low enough to produce a sufficient accumulation of net damage to account for the observation that atherosclerosis and long-term blood pressure increases with age occur in very large numbers of "normal" people.

Second, the basic concept of these diseases as accumulations of many small damage events implies that the number of damage events accumulated by different individuals will show a broad continuous distribution. Because of this, any agent that accelerates the production or prevents the repair of such damage events will shift all, or a very large part, of the exposed population in the direction of worse function. For example, noise exposure will cause people who would otherwise have excellent hearing to have less excellent hearing; people who might have fair hearing without noise can expect to be shifted toward poor hearing, and so forth.

Many years ago, Weiss and Spyker (1974) suggested that methylmercury might accelerate the loss of neurons in adult life and contribute to a "chronic cumulative" process that would fall within this category. Recently, an apparently very sound case control epidemiological study among people in Singapore found a strong association between blood levels of mercury and the risk of Parkinson's disease (Ngim and Devathasan, 1989). Parkinson's disease occurs when a large portion of the neurons responsible for making dopamine in the substantia nigra area of the brain is lost. Thus, if the association with mercury exposure holds and proves causal in later studies, it would be a very important basis on which to quantitatively reevaluate the long-term risks of seafood mercury consumption.

CRITIQUE OF RISK ASSESSMENTS

One of the charges to the committee was to review and critique previous agency efforts to assess the likely human risks of chemical contaminants of aquatic organisms. The committee was asked to suggest directions for the improvement of risk assessment approaches to better serve the needs of decision making on chemical risks, but not to attempt to develop its own set of final risk numbers. The committee has chosen to focus on PCBs and methylmercury as two paradigmatic cases, representing one carcinogen and one noncarcinogen, for which there is a relatively sufficient public record of FDA efforts to assess the risks and benefits of alternative control options.

The PCBs are of special interest in light of the results presented in this chapter. By using conventional carcinogenic risk assessment techniques (although these have considerable difficulties in general and especially as applied to PCBs) and available data on likely average daily intakes in the United States, PCBs are shown to pose by far the largest *potential* carcinogenic risk of any environmental contaminants for which measurements exist. According to EPA's cancer potency factor, the aggregate U.S. lifetime risk could be as high as 6×10^{-5}, corresponding to approximately 120 cancers per year. With FDA's lower estimate of the upper confidence limit (UCL) of the PCB cancer potency, the aggregate risk for the U.S. population would be considerably less, but still not insignificant, 2.7×10^{-6}, implying a maximum toll from commercial seafood of about 5 cancers per year.

These risks would be much greater for the subpopulation consuming relatively large amounts of sport or subsistence caught fish near the FDA tolerance level of 2 parts per million (ppm). For someone who consumes 20 kilograms (kg) per year of 20-ppm fish [which is probably not far from the actual consumption of some izndividuals in some areas of Lake Michigan (Humphrey, 1983a,b)], the upper-confidence-limit lifetime risk that would be expected by using the FDA cancer potency estimates is about 5×10^{-4}; for the EPA cancer potency estimate, the lifetime upper-confidence-limit risk would be somewhat more than 1%. Perhaps in part because of the difference in cancer potency factors, there is also a great difference in the practical policies of the two agencies in regulating PCBs, with FDA generally taking a far less protective posture toward PCB exposure than EPA.[6]

Methylmercury is also of special interest because it shows the operation of the old no-effect level/safety factor paradigm in a practical case where, as it happens, an alternative type of quantitative assessment is also possible based on existing data. In this case, therefore, one can examine what levels of risk might implicitly be accepted by using FDA's rule-of-thumb procedure for dealing with noncancer effects.

In the following material, the committee will assess for both casesthe reasonableness of the calculations that went into FDA's decision making in terms of the information available at the time; andthe improvements that could and should be made by applying more modern analytical techniques to recently available or feasibly obtainable data on population exposures, the nature and mechanisms of potential adverse effects, and likely population dose-response relationships.

To briefly foreshadow the committee's conclusions, three statements can be made:

1. Both assessments (Cordle, 1983; Cordle et al., 1982; FDA, 1979; Tollefson and Cordle, 1986), even at the times they were done, suffered from ad hoc unsystematic approaches to the treatment of population heterogeneity in dosage and (for methylmercury) susceptibility to adverse effects. For mercury, the likely special susceptibility of developing fetuses was mentioned in discussion. However, a tenfold "safety factor" was applied to the lowest blood level reported to produce effects for adults (rather than to a no-effect level of intake, which would have been more consistent with established procedures) in the cited epidemiological studies, without numerical allowance for extra sensitivity of fetuses, without a quantitative dose-response treatment of the data then available for adults to gauge the potential adult risk at the blood level selected as the highest permissible for U.S. consumers, and without a quantitative treatment of the effects of pharmacokinetic differences among people that would tend to make blood levels and risks as a function of dietary intake more variable than blood levels themselves.
2. Procedures for the systematic quantitative analysis and communication of uncertainties are absent. Both analyses use the "method" of compounding a series of ad hoc allegedly "conservative" assumptions (such as the use of the consumption level of the 90th percentile consumer, the upper 99% confidence limit for the PCB animal potency calculation, but a body weight rather than a surface area scaling rule to translate animal to human dose). The difficulty in picking a single series of point estimates of uncertain parameter values is that after the first few such assumptions have been entered into an analysis, no person on earth can determine where one actually is on an overall distribution of the likelihoods of different outcomes. Monte Carlo simulation procedures are now readily available via personal

computer-based software for the calculation of probability density functions of different outcomes, given uncertainties in multiple parameters affecting population risk. Such procedures have been effectively advanced in the context of risk assessment in a recent report by Finkel (1990).

3. Both assessments have been rendered substantially obsolete by the development of recent information related to PCB and to methylmercury risks, changes in the economic conditions assumed in the cost analyses, and possibly changes in residue levels and available options for population exposure reduction.

In the case of PCBs, understanding of mechanisms has advanced to the point where there are now serious proposals for a congener-specific assessment of relative activity (Clement Associates, 1989; Jones, 1988). In the view of the committee, this congener-specific activity analysis must be coupled with a congener-specific treatment of pharmacokinetics, which now seems possible based in part on the human dietary exposure and serum concentration data being collected by Humphrey and coworkers (H. Humphrey, Michigan Department of Public Health, Lansing, personal communication, 1989). Both congener-specific activity estimates and pharmacokinetics should be used to evaluate not only existing animal carcinogenesis data, but also recently emerging epidemiological information from studies in workers.

In the case of methylmercury, recent data allow preliminary evaluation of the population dose-response relationship for a number of long-term neurological effects from exposure during fetal development (Marsh et al., 1987). The analysis in the following section indicates that the apparent interindividual variability in susceptibility for fetal effects is much broader than that for adult effects. Consequently, although the tenfold safety factor, as applied, appears to offer a reasonable degree of protection for adult effects, projections based on an additional toxicological assumption of lognormal distribution of threshold for effects (logprobit projections) of the fetal dose-response data suggest the possibility of appreciable risk from methylmercury exposure, even at levels to which many people are exposed via the diet. Published projections applying a logistic model to the same data come to similar conclusions (Cox et al., 1989).

Polychlorinated Biphenyls (PCBs)

Background

The PCBs originally came to the attention of FDA through a series of contamination incidents in which PCB-containing oils leaked from electrical equipment directly into food substances, such as rice oil in Japan and Taiwan (the "Yusho" incidents), and by various routes found their way into U.S. livestock. The initial concerns were direct overt toxicity and reproductive effects, and acceptable daily intakes were defined as 100-fold and 10-fold below the no-effect levels in animals and humans, respectively.

To this day, FDA analyses continue to reflect and even emphasize calculations within the NOEL/safety factor paradigm that was initially used to interpret animal and human data on the risk of overt chronic toxicity. Thus, Scheuplein (1988) compares FDA's 1 microgram (μg)/kg/day ADI to the observations of Humphrey (1983a) that average PCB intakes in a group of heavy eaters of fish[8] from Lake Michigan were 1.75 μg/kg/day:

> No symptoms or adverse health effects clearly related to PCB ingestion could be identified in the exposed group. This suggests that FDA's acceptable intake level of 1.0 μg/kg-day provides reasonable safety, although it is possible that effects too subtle for detection are occurring or that the latency periods for cancer are very long.

From work reported to EPA (Humphrey, 1983b) and other data reported later by Humphrey 91988), it appears that the cited work had only a very small health component (major emphasis was on defining exposures).

The health component seems to have been a simple medical history questionnaire administered to the participants, asking them about 18 specific conditions. This limited presentation of the data, without confidence limits or sample sizes, may not do justice to the underlying work. Clearly, however, this kind of study was not designed to be a very sensitive detector of carcinogenic risk or other types of effects (e.g., on fetal development) that are the concerns at low doses. Carcinogenic risks are traditionally studied in human populations by case control studies and detailed comparisons of the frequency of specific cancers or cancer deaths in defined cohorts, adjusted for age.

Reproductive effects must clearly involve careful follow-up of offspring in groups where – through either dietary analysis or serum studies– it is possible to define prenatal exposures. The fact that the Humphrey (1983a) work is seriously cited as evidence of no effect – and that other work (e.g., Fein et al., 1984; Jacobson et al., 1985) that does suggest fetal effects from the same general Michigan fish eater population is not cited – is perhaps testimony to FDA's general lack of enthusiasm at the prospect of departing from the NOEL/safety factor paradigm.

FDA Assessment of Costs and Risks for PCBs in Fish

The most recent detailed statements concerning FDA assessment of the carcinogenic risks of PCBs appear to be in two papers by Cordle (1983) and Cordle et al. (1982). These recapitulate and give more details of the analysis underlying the official risk assessment that appeared in the *Federal Register* (FDA, 1979) promulgating the 2-ppm tolerance (lower than the former 5-ppm guideline). This analysis appears to have been done in part to satisfy requirements for a balancing analysis of benefits and costs that became a government-wide requirement for regulatory action in the mid-1970s. For this purpose, however, the calculation of societal aggregate cancer risks from 90th percentile values of consumption for fish consumers is problematic because, of course, the 90th percentile overstates the average. This probably results in a small error in the opposite direction, because of the likely skewed distribution of PCB consumption among fish consumers.[9] The committee's aggregate national average estimate of current PCB exposure from fish is slightly less than 0.01 μg/kg/day.

Had it done so, FDA might at least have been led to some explicit comment on the closeness (within twofold) of the incremental cost/benefit ratio of the regulatory step it declined to take (from 2 to 1 ppm) to the incremental cost/benefit ratio of the regulatory step it did take (from 5 to 2 ppm). If FDA really intended this result, then an unusually specific insight is afforded into the value it attached implicitly to the prevention of a unit of upper-confidence-limit cancer risk.

Instead, it appears that the FDA decision was most influenced by the concentration of the regulatory impact expected for commercial harvesting of freshwater fish. Whereas the 2-ppm tolerance was expected to cause the loss of a negligible portion of the commercial marine catch (less than 0.2%), the same regulation was expected to cause the loss of 14% of the freshwater catch. For the 1-ppm standard, the loss of freshwater fish was expected to increase to 35%. In its conclusion commenting on the choice of the 2-ppm level, FDA focuses on the absolute magnitude of the losses expected from the 2-ppm and 1-ppm alternatives:

... the agency estimates that under a tolerance of 1 ppm, approximately $16 million worth of the commercial fish catch would be violative and thus, presumably, removed from commerce. This is nearly triple the $5.7 million worth estimated to be violative under a 2 ppm tolerance. It is far more likely under a 1 ppm tolerance than under a 2 ppm tolerance that the more heavily contaminated species of freshwater fish would be violative in percentages high enough to put an end to their commercial exploitation and, possibly, force some segments of the freshwater fishing industry to cease operations completely.

Thus, the actual loss of food resulting from the 1 ppm tolerance could greatly exceed even the $16 million landed value (1974 dollars) estimated above. Second, for the average consumer, current exposure to PCBs in fish is at a tolerably low level, when considered in the light of the criteria of section 406 of the act [Food, Drug, and Cosmetic Act], without a 1 ppm tolerance. The average consumer eats a modest amount of fish from a variety of sources, both freshwater and marine, most of which yield fish with PCB levels below 1 ppm. Because their exposure is thus low to begin with, they are adequately protected by a 2 ppm tolerance, which ensures that they will not be exposed to the unusually high levels of PCBs found in some species of fish. The slight additional protection these average consumers might gain from a 1 ppm tolerance does not justify the significantly greater impact such a tolerance would have on the availability of food. On the other hand, atypical heavy consumers (e.g., the Great Lakes sport fisher who catches and consumes large quantities of the contaminated species) would likely not be adequately protected by even a 1 ppm tolerance because of the amount of fish they eat and because those fish are seldom affected by FDA tolerances (either because they are sport fish or are from intrastate commercial channels and, in either case, are outside FDA's jurisdiction).

Some additional portion of the difference derives from the fact that EPA has used more recent experiments with Aroclor 1260 for its risk calculations (Norback and Weltman, 1985), whereas FDA used data on the overall tumor risk from an experiment with Aroclor 1254, which produced the highest estimate of risk from studies then available (Cordle et al., 1982).

In the context of the carcinogenic risk assessment practices of its time, the FDA quantitative analysis does not greatly depart from established principles. On the theory (discussed above) that cancer is the result of a multiple series of somatic mutations, a hypothesis of low-dose linearity in carcinogenic response is not at all unreasonable, particularly if carcinogenic transitions caused by one agent can interact with some of the processes that cause the vast number of "background" cancers in the human population. An upper-confidence-limit risk derived from the multistage theory guarantees low-dose linearity in risk estimates and thus provides a convenient benchmark for risk even though a "mean-estimate."

One can question the practice of summing up tumors from all sites in terms of the multistage theory, but this is not unusual in the context of risk assessment practices in the late 1970s and early 1980s. The choice of animal to human extrapolation formulas is open to considerable discussion in the field to this day, although in the specific case of PCBs, from some limited positive epidemiological studies in PCB-exposed workers, Allen et al. (1987) have estimated a TD_{25} (the lifetime dose estimated to produce cancers in 25% of exposed people) of 0.15 mg/kg/day (for 45-year, 240-day/year occupational exposures). This translates into the equivalent of a best-estimate cancer potency factor of 3.9 cases per lifetime mg/kg/day dosage– rather closer to the EPA upper confidence limit estimate of 7.7 than to the FDA estimate of 0.34 case per lifetime mg/kg/day.

Overall, FDA (1979) has been quite cautious in presenting the difficulties and uncertainties of carcinogenic risk analysis for PCBs from the data it had available: ... the utility of this risk assessment for evaluating actual risk to humans from exposure to PCBs is extremely limited. This is due both to difficulties inherent in making such extrapolations from animals to humans and, perhaps more importantly in this instance, to gaps and uncertainties in the data available for this particular risk assessment. For example, the toxicity studies on which the risk assessment is based used commercial preparations of PCBs, which are chemically different from the PCB residues found in fish and which contain small amounts of highly toxic impurities (e.g., dibenzofurans) not known to be present in fish residues. [It should be noted, however, that these contaminants are also not known *not* to be present in fish.] Also, in making the exposure estimates required for the risk assessment, it was necessary to use existing data on the numerical distribution of PCB levels in fish and rely on the assumption that the effect of a given tolerance level is to remove from commerce all fish containing PCBs exceeding the tolerances. It is possible that neither the assumption nor the data precisely reflect what actually occurs.

For these reasons and others ... the risk assessment does not provide a basis for precise quantification of the amount of risk reduction accomplished by reducing the fish tolerance. Despite the limitations inherent in the risk assessment, however, the agency regards it as illustrative of the basic validity of the toxicological rationale for reducing the tolerance for PCBs in fish: Reduction of the tolerance will result in a significant reduction in risk among those who consume PCB-contaminated fish. FDA considers this risk reduction to be of significant public health value, even though it cannot be precisely quantified. For this limited purpose, the analysis as presented may well have served an appropriate and useful function. However, there are now both an opportunity and a need to gather and analyze newer information, so that by the early to mid-1990s it will be possible to significantly improve on this 1979 effort.

Suggestions for Improved Analysis

First, a better regulatory impact analysis might evaluate *a wider range of choices* of regulatory alternatives–including not just different tolerance levels, but possible rules restricting the location of harvest, species, and size. Choices in these areas are likely to allow a more narrowly targeted regulatory action that will reduce human exposure to PCBs at less overall cost of forgone food resources. By being keyed to characteristics of fish that do not require expensive and time-consuming chemical analyses, such regulations might be much more efficiently implementable than tolerances based simply on chemical content.

Second, an improved analysis of carcinogenic risks should be based on a congener-specific assessment of (1) pharmacokinetics and (2) relative potency at the site(s) of action. The FDA analysis mentions the fact that the distribution of PCB congeners to which consumers are exposed is not the same as found in the commercial PCB preparations (e.g., Aroclor 1254) that were the subjects of carcinogenicity testing. The FDA specifically mentions the possibility of dibenzofuran contamination in the commercial mixtures used for carcinogenicity testing. Not mentioned, however, is the likely possibility that fish PCB residues may be relatively enriched in just those congeners that are more persistent in biological systems. More degradable congeners are almost by definition likely to be preferentially destroyed either in the fish themselves or in organisms that are lower on the food chain, whereas the original commercial mixtures would contain the full mix of more- and less-degradable congeners. Humphrey (1983b) noted a change over time in the distribution of PCB congeners found in human serum toward more highly chlorinated congeners, which tend to be more persistent in biological systems. Data now exist that would allow a detailed comparison of congener distributions in fish and humans with those in commercial mixtures (McFarland and Clarke, 1989). At the same time, a long series of comparative studies by Safe (1989) and others has provided basic insights into the mechanisms of action of PCBs, dibenzofurans, and dioxins. It now appears likely that PCBs enhance carcinogenesis not by direct interaction with DNA, but through a receptor-binding mechanism (similar to dioxins) whose precise dose-response implications have not been elucidated. Receptor binding and related activities vary enormously among different congeners, with congeners that are planar in shape (and relatively highly chlorinated) appearing to have the greatest potency. Careful experiments relating tumor enhancement to the number, type, and persistence of specific PCB-receptor interactions in animal systems could yield important insights into the likely form of the dose-response relationships that should be applied to human risk assessment. There are now serious proposals for a congener-specific assessment of relative activity by using the existing data base (Clement Associates, 1989; Jones, 1988). This approach needs to be pursued and refined with additional laboratory and carcinogenesis studies.

A good opportunity may exist to reevaluate the accumulating evidence of human carcinogenic response in occupationally exposed populations with relation to serum PCB levels and to relate the human response per unit of delivered dose to the response per serum level likely to have been present in rodent bioassays. [There are two very limited but apparently positive epidemiological studies in groups of capacitor workers (Bertazzi et al., 1987; Brown, 1987) and at least one negative study at a plant in Massachusetts whose results have not yet been published in detail. In at least one case, there are measurements of serum PCB levels in the exposed group.] Harold Humphrey of the Michigan Department of Health reportedly has an extensive data set of PCB concentrations in blood in relation to dietary exposure levels. A series of congener-specific measurements is now in progress together with published measurements of metabolic rates of PCBs in humans (Buehler et al., 1988; Phillips et al., 1989; Yakushiji et al., 1984); this is likely to be useful in establishing metabolic rates of PCBs for comparison with animal models and measurements after the work by Matthews and Dedrick (1984) and Tanabe et al. (1981).

Animal work, together with the newer metabolism information, appears capable of providing a basis for estimating serum PCB levels in the rats and mice that showed positive cancer responses in chronic bioassays. Projections to human risk could then be made on the basis of a more appropriate measure of "delivered dose" (weighted for different congeners by receptor binding and pharmacokinetics) rather than gross dietary intake. There already is reasonably good information on the population distribution of overall plasma PCB levels in the United States Using such a distribution, if one had an estimate of cancer risk per unit of plasma PCB concentration, an inference could be made not only about the average risk to the population, but about the distribution of risk to various individuals that derives from differences in both dietary habits and PCB elimination rates. Plasma and whole body burden PCB levels could also be productively used as indicators of long-term dosage in epidemiological case control studies of cancer patients (versus other people) in communities near the Great Lakes with greater than usual numbers of people having high exposures to fish with appreciable levels of PCBs.

A better analysis should also include a quantitative assessment of the risks posed by prenatal exposure to PCBs, via such parameters as gestational period and birth weight, as well as subtle neurological measures of effect. The presence of at least modest effects on birth weight has recently been supported by an epidemiological study in workers (Taylor et al., 1989). Because birth weight is such an easy parameter to study (measurements are made routinely on nearly all newborns) and because it has a very strong relationship to infant mortality (Hogue et al., 1987), additional large population studies of birth weight and gestational period in relation to fish consumption and relevant serum PCB congener levels should be a high priority. Methodology exists to interpret birth weight changes in terms of other relevant endpoints such as infant mortality (Ballew and Hattis, 1989), but these techniques require validation with an actual case having large enough numbers of subjects to provide data on both the intermediate parameters (birth weight and gestational age) and the ultimate parameters of concern (infant mortality, impairment of neurological development, etc.). Finally, there is a need for improved and more systematic treatment of analytical uncertainties, with ultimate results in the form of cumulative probability density functions or, failing this, at least a statement of both best-estimate and upper-confidence-limit risks, under consistent and defensible theories.

Methylmercury

FDA Risk Assessment and Current Regulations

The published analysis of Tollefson and Cordle (1986) appears to be the most definitive public assessment by FDA personnel of the risks of methylmercury in fish.

The principal benchmark that Tollefson and Cordle (1986) use in defining an acceptable intake of methylmercury is the lowest blood methylmercury level that appeared to be associated with overt neurological symptoms of toxicity in a few studied mass poisoning incidents (two that occurred from fish taken from Minamata Bay and Niigata in Japan in the 1950s and 1960s, and one incident that arose from the consumption of contaminated seed grain in Iraq): A Swedish expert group evaluated data on human methylmercury toxicity derived from cases which occurred in Minamata and Niigata. The Swedish group determined (by extrapolation) that the lowest blood mercury level associated with toxic effects was 220 ppb [parts per billion] and the lowest hair mercury level associated with toxic effects was 50 ppm (Berglund et al., 1971).

The Iraqi outbreak of methylmercury poisoning has been extensively studied by many investigators. The blood level of mercury at which symptoms of toxicity were first detected in the Iraqi episode was approximately 240 ppb [parts per billion] (Clarkson et al., 1976). This calculation was made on samples collected 65 days after the end of exposure, which is the approximate half-life of methylmercury in humans. Since the actual clearance times from the blood are not known, the level may lie between 240 and 480 ppb. These values are for adult exposures only. Studies of people in Sweden who consumed relatively large amounts of fish allowed these blood levels to be related to dietary levels–at least on a population average basis (Tollefson and Cordle, 1986): "A linear relationship was found between daily ingested methylmercury and the level of methylmercury in blood, and the data indicated that a steady daily intake of approximately 300 μg Hg [mercury] as methylmercury for a 70-kg person would result in a blood concentration of roughly 200 ppb at steady state (Skerfving, 1974)."

This 300-μg/day dose rate, calculated to correspond to 200 ppb in blood *for an average person*, was interpreted by Tollefson and Cordle (1986) as a threshold value at which symptoms of toxicity associated with methylmercury are first noticeable. ... Dose-response relationships below this range of intake are not known. In addition, there is concern about the relative sensitivity of the developing fetus. The question of interaction of other chemical factors such as selenium on the toxicity of methylmercury has not been conclusively demonstrated at this time, but may be a factor to be considered when more information is developed. Because of these areas of uncertainty, a safety factor of ten has been used to provide a sufficient margin of safety.

Thus a maximum tolerable level would be 30 μg methylmercury daily in the diet, resulting in 20 ppb of methylmercury in blood and 5 ppm in hair.

The following limitations to this approach were recognized:

(1) it was not known to what extent particular individuals are more or less sensitive to mercury than others;

(2) the estimates were based on the "lowest level that caused an effect" rather than the normal procedure of using a "no-effect dose level";

(3) paresthesia is usually the first symptom of methylmercury toxicity noted but is not sufficient to diagnose poisoning because it can be caused by many other factors (Clarkson et al., 1983);

(4) questions about dose-response relationships in human fetuses and newborn infants were unanswered; and

(5) there is a possibility of subclinical effects arising from exposure to very low levels of methylmercury.

Tollefson and Cordle (1986) therefore realize that they are departing from the usual procedure for the NOEL/safety factor analysis and that, in particular, risks to fetuses may not be represented adequately in the available data. However, they make no additional numerical adjustment for these uncertainties, presuming implicitly that the usual 10-fold safety factor has enough of a margin built into it to avoid an unacceptable level of risk.

In addition to these enumerated factors, there is a subtle point that makes the calculation on the basis of the average diet/blood relationship less protective than the usual calculation based on a 10-fold reduction from the dietary dose that is directly observed to be without effects. The 10-fold factor is supposed to represent the protection needed to guard against interindividual variability in the complete pathway from dietary intake to the production of biological responses. Some portion of that human interindividual variability, however, occurs in characteristics that affect individual blood levels–especially individual half-lives for elimination of methylmercury from the body. For those with relatively slow elimination rates, toxicity would be expected to be associated with lower average dietary intakes than for those with average elimination rates. Therefore, by doing the dose extrapolation in terms of blood levels, and translating the corresponding *average* dietary levels, the authors are being less protective than would have been the case had the 10-fold factor been applied to the minimum dietary dosage associated with an observed effect (or no observed effect). Later in the paper, Tollefson and Cordle (1986) discuss interindividual differences in biological half-lives for methylmercury as follows: Additional data developed on the biological half-life of methylmercury in humans, however, indicate a need to take into account the problem of variations among individuals. In the Iraqi episode, 90% of the individuals studied had a biological half-life of methylmercury between 60 and 70 days, but 10% showed values of 110 to 120 days (Al-Shahristani and Shihab, 1974). Individuals having a long biological half-life would accumulate much higher steady-state levels than those having short biological half-lives and would thus be at greater risk from the same level of methylmercury intake.

Tollefson and Cordle have unfortunately misread the population distribution of methylmercury half-lives presented in the cited paper. A figure in the paper is misleading in that the highest single observed biological half-life (189 days) has somehow been excluded, but there is no way that even the figure can be construed to indicate that 90% of the values are between 60 and 70 days.

Tollefson and Cordle (1986) go on to support the results of their analysis by citing studies failing to find obvious methylmercury toxicity in heavy fish-eating population groups in Peru and American Samoa with average blood methylmercury concentrations in the range of 60-80 ppb (Marsh et al., 1974; Turner et al., 1980). However, without careful analysis of the distributions of thresholds implied by the earlier data, the sample sizes, and the degree of background symptoms in control groups in the negative studies cited, neither the reader nor the authors can judge the sensitivity of those studies to detect adverse effects and whatever quantitative implications there might be for the level of risk at lower doses.

Tollefson and Cordle's apparent reluctance to construct and analyze the implications of quantitative population distributions is also seen in their treatment of the expected dietary dosage of methylmercury to consumers. They had available the results of a fairly extensive study commissioned by the Tuna Research Foundation (TRF) of one-month diary reports of consumption of different types of fish by 7,662 families, including 25,165 people, and measurements of mercury levels in different species of fish. However, instead of using these data to describe the population distribution of monthly mercury intakes from fish as a whole and the possible effects of different tolerances, Tollefson and Cordle (1986) opted for an ad hoc treatment of the data for only a few species of interest: Data from the TRF survey have indicated that the average consumption of all fish among the fish-eating population of the United States is 18.58 oz/month or approximately 18 g/day. Daily consumption of species containing relatively high levels of methylmercury such as tuna, swordfish, or halibut would be considerably less. For example, the mean daily consumption of halibut is 7.2 g with two standard deviations increasing the total to 16.6 g. This would include some 97.5% of all consumers of halibut. The consumption of 16.6 g of halibut with 0.179 ppm mercury would provide a daily mercury intake of approximately 2.9 μg.

The mean daily consumption of swordfish is 6.53 g with a standard deviation of 2.5 g. If 11.53 g of swordfish with a mercury level of 1.5 ppm were consumed each day, and this would include over 95% of all swordfish eaters, the daily mercury intake would be 17.3 μg, still below the ADI of 30 ¼g.

At the highest level of swordfish consumption shown by the NPD [National Purchase Dairy Panel, Inc.] study, i.e. 511 g in a month or 17 g per day, with a mercury residue of 1.5 ppm the daily mercury intake would be 25.5 μg, still below the ADI of 30 μg. If in addition to the highest level of consumption of swordfish in the NMFS [National Marine Fisheries Service] study, i.e. 17 g per day with a mercury residue of 1.5 ppm, the same individual consumed the average daily amounts of tuna, halibut, and salmon at the present average residue levels of mercury for each of the three species, then the total daily intake of mercury from all four species would average 31.7 μg, only 1.7 μg above the ADI.

Such consumption, which would realistically be on a periodic basis to derive daily exposure, seems very unlikely, particularly since the cost of swordfish, halibut, and salmon is prohibitively high. Tollefson and Cordle are certainly correct in concluding that very few consumers will exceed the 30-μg/day level for prolonged periods. However, this type of analysis cannot give an estimate of how many might, or how many might fall short of the 30-μg/day level by various amounts. The failure to analyze the data more rigorously is all the more unfortunate because the FDA had available an example providing statistically sophisticated treatment of these same data (NMFS, 1978).

Tollefson and Cordle (1986) conclude with a somewhat unusual statement of the relevant risk issues: The data currently available for evaluating quantitatively the association of neurological symptoms of toxicity with exposure to methylmercury are sparse and inconsistent. Additional studies are being carried out on the prenatal effects of methylmercury to determine that this lifestage continues to be protected by the 1.0 ppm regulatory level for mercury in fish. However, even with the above-outlined uncertainties concerning the results of exposure in Japan and Iraq, where exposures were considerably higher than anything experienced in other countries, U.S. fish consumption data do not indicate any cause for concern of methylmercury poisoning for the *average American.*

The majority of fish consumers could easily double their intake and still remain below the mercury ADI. The current 1.0 ppm regulatory level for marine species provides more than adequate protection *at the current average fish consumption level in the U.S.* In addition, the enforced limit of 1.0 ppm mercury in marine fish provides a sufficient margin of safety for young children and for significant numbers of consumers exceeding the acceptable daily intake. The potential for a more quantitative analysis of methylmercury risks with the aid of available data on the distributions of blood methylmercury concentrations and dose-response data from the Iraqi poisoning incident is illustrated below. The aim is not to provide a definitive assessment of methylmercury risks, but to show an approach for improving the type of assessment now done by FDA.

Available Data on Blood Levels in the U.S. Population

Gathering and analyzing data on the actual distribution of blood mercury and methylmercury levels in the population allow one to bypass a great deal of the complexity of modeling individual dietary intakes over times long enough to determine biologically significant systemic exposures. Full data describing this distribution are not available; however, a plot of the median and extreme values given indicates that the distribution is well described as lognormal. The variability indicated by the $\log_{10}$ geometric standard deviation of 0.328 indicates that 95% of the population is spread out over a 20-fold range – between 4.5-fold above and 4.5-fold below the median blood level for the population. This variability includes both dietary and pharmacokinetic differences among individuals.

Algonac (population 3,684) was chosen because of its proximity to both Lake St. Clair and the St. Clair River – sites of a famous mercury contamination episode discovered by a Canadian graduate student in 1970. South Haven (population

6,471) was chosen as a control community with similar proximity to Lake Michigan. Recently, Marsh et al. (1987) published detailed information on the incidence of a variety of fetal mercury effects in relation to the maximal levels of mercury found in the hair of the mothers during gestation. (The observations come from an Iraqi mass poisoning incident that resulted from the distribution of methylmercury-treated Green Revolution seed grain.) In all, 81 mother-infant pairs were studied. Such a conclusion, however, depends on whether the biomarker of exposure used in this case – the maximum hair mercury found at any time during gestation – is the most appropriate direct causal predictor of response that can be developed. Other possibilities might well include the concentration of mercury at a specific sensitive time during gestation or a weighted sum of concentration x duration over a specific set of sensitive periods. Accurate assessment of the degree of interindividual variability in susceptibility in humans, and consequent low-dose risks, might well depend on quantitative measurement and modeling of the causal processes involved in this case and reanalysis of the data according to the most likely causally predictive summary measure of delivered dose.

It can be seen that, as with fetal effects, the fuller analysis of data with the Finney (1971) procedure suggests somewhat larger probit slopes than the values determined by simple regression analysis of the points. It can also be seen that, in comparison with the fetal effects, the effects in adults tend to have larger probit slopes, indicating less interindividual variability and therefore less risk at low doses.

Biologically, the implication is that the more complicated array of causal processes involved in producing fetal effects (allowing for variability in maternal mercury elimination, maternal/fetal transfer, and the time-sensitive processes of interference by methylmercury with development) tends to lead to more interindividual variability. From a social policy perspective, the implication is that the 10-fold safety factor rule, whatever its effectiveness in protecting exposed populations against the effects of interindividual variability in adults, appears likely to provide less protection for the developing organism.

Projections of fetal risks here, of course, are greatly affected by the assumed lognormal form of the distribution of thresholds for both adult and fetal effects. Although lognormal distributions appear to be broadly compatible with the high-dose data (see earlier figures), this is not the only statistical form that the distribution of thresholds might take. Cox et al. (1989) have applied a logit model, which assumes a different distribution of thresholds, to predict possible low-dose risks for one of the fetal effects covered here (late walking), using the same data set (Cox and coworkers also made projections based on scores for neurological impairment, but using a different cutoff for defining effect than the committee, thus rendering the risk projections noncomparable).

As mentioned earlier, it would be more common for a 10-fold safety factor to be applied to the dietary dosage of a toxicant, rather than to blood levels. To represent this for methylmercury, one must add to the logprobit risk equations an estimate of the interindividual variability in internal blood levels as a function of dietary intake. In principle, two factors should be considered in accomplishing this: (1) interindividual variability in absorption, and (2) interindividual variability in overall elimination of mercury from the blood. Methylmercury is relatively efficiently absorbed from food, so the influence of the first factor may well be small. The effect of this is to increase the estimate of the probit slope by a relatively greater amount for those effects that, in the absence of the adjustment, had relatively small amounts of interindividual variability (high probit slopes). The corresponding estimates of adult risks are increased, but generally not by a huge factor, for the adult effects expected to pose the numerically greatest risks.

Conclusions and Recommendations for Changes in Risk Assessment Practices

As indicated even by these two examples (PCBs and methylmercury), where FDA has devoted much more attention to two types of residues than to other potential chemical hazards of seafood consumption, the agency lags badly in the development of innovative methodology for assessing risks and evaluating the potential benefits of different options for control. This is all the more unfortunate because FDA has major technical resources available in both risk assessment and experimental toxicology (including reproductive toxicology) at the National Center for Toxicological Research in Arkansas. Neither the PCB nor the methylmercury analyses by FDA show evidence of input from this very capable group, which is under FDA's jurisdiction.

Carcinogenesis risk assessment procedures should be modified to give decision makers additional information about the uncertainties of the analysis, as well as both societal aggregate and individual risk estimates. In particular, to facilitate evaluation of the costs and benefits of measures to achieve quantitative reductions in exposure to carcinogenic inadvertent contaminants, procedures should be developed to supplement current "upper-confidence-limit" cancer potency estimates with estimates representing the central tendency of cancer risks, information on cancer risks from all available species, and comparative information on the pharmacokinetic and pharmacodynamic factors in different species.

6

Technical Problems in Upkeeping

The recent introduction of several transposable elements in zebrafish opens new frontiers for genetic manipulation in this important vertebrate model. This review discusses transposable elements as mutagenesis tools for fish functional genomics. We review various mutagenesis strategies that were previously applied in other genetic models, such as Drosophila, Arabidopsis, and mouse, that may be beneficial if applied in fish. We also discuss the forthcoming challenges of high-throughput functional genomics in fish.

INTRODUCTION

Zebrafish exhibits many qualities that are essential in a good genetic model: relatively small size for vertebrates, high fecundity, reasonably short generation time, low maintenance costs, amenability to large-scale saturating mutagenesis screening, availability of mutants, genetic maps, and a sequenced genome. However, the lack of some classical tools in genome manipulation has been a serious drawback. In recent years, several transposable elements from heterologous hosts (Tc3, mariner, Tol2, Sleeping Beauty, Frog Prince, and Ac/Ds) have been successfully applied in zebrafish. All of these 'cut-and-paste' transposon vectors are designed to operate as two component systems: a non-autonomous element that inserts into the genome of the host and a transposase that catalyzes transposition of the non-autonomous element. A non-autonomous element can carry a cargo inserted between the two end sequences that are required in cis for transposition. It is immobile in the absence of its transposase. The transposase is usually supplied in the form of mRNA by microinjection into fertilized eggs to induce transposition of the non-autonomous element. The ease of microinjection in fish is advantageous; the injected transposase mRNA is degraded some hours after injection, and new insertions of the non-autonomous element in the genome remain immobilized.

The choice of many transposon vectors provides zebrafish biologists with a plethora of possibilities, but it also raises the issue of which to use and when. The Tc3 element from Caenorrhabditis elegans and the mariner (Mos1) element from Drosophila mauritiana were the first transposon vectors used to generate transgenic fish. However, since the initial reports, no further developments using these transposons have been described in the literature. Two transposable elements of fish origin, Tol2 from medaka (a member of the hAT [hobo/Ac/Tam3] superfamily) and Sleeping Beauty (a synthetic member of the Tc1/mariner superfamily), have both become widely used in the past few years. Recently, a transposable system from plants has been added to this arsenal. The modified maize Ac/Ds elements are effective transformation vectors and produce high genomic Ds transposition and re-transposition rates in zebrafish.

From the available reports, it appears that Sleeping Beauty produces lower transgenesis rates and is more sensitive to size of cargo fragments compared with Tol2 and Ac/Ds. Nevertheless, it should be possible to transfer large transgenes using a modified strategy. The Tol2 and Ac/Ds systems produce the highest transgenesis rates reported to date, and they are capable of carrying large cargo fragments without significant reduction in activity. Both systems can generate multiple insertions that are advantageous for insertional mutagenesis. On the other hand, working with multiple insertion lines and maintaining them can be cumbersome. Furthermore, highly active transposons can undergo multiple excision-insertion events, leaving behind the untagged footprints and genomic rearrangements. It is surprising that despite a huge phylogenetic distance between hosts, the Ac/Ds system performs as well in zebrafish as in plants. Moreover, transposition activity of Ac/

Ds in fish is comparable to that of the fish-derived member of the hAT superfamily Tol2. The latter system has successfully been tested in several distant vertebrate hosts. The Ac/Ds system has been utilized in many plant species, yeast, and now in vertebrates, which is the widest host range demonstrated for any transposon. These facts strongly suggest that members of the hAT superfamily are probably the most versatile cut-and-paste transposons, and can be broadly utilized in genetic manipulation of various organisms. This makes hAT elements the best candidates for introduction into new species that do not yet have established transgenesis and insertional mutagenesis techniques.

The availability of alternative transposon systems is beneficial for various reasons. Different insertion preferences produced by different transposable elements can allow better distribution of insertions throughout the genome. The absence of cross-activation between different transposons permits independent sequential use of several elements within the same animal and can be useful for various applications. For example, one can envision using one transposon to generate transgenic reporter lines for any given cell type of interest, and these lines can subsequently be mutagenized using a different transposon to create a bank of mutations affecting that particular cell type.

The simplicity and the ease of use, and the ability to carry complex cargo fragments, combined with high cargo capacity, makes transposons superior to retroviral vectors for fish transgenesis. Thus, generating transgenic fish using transposable elements is becoming a routine laboratory technique.

MUTAGENICITY OF TRANSPOSON INSERTIONS

Vertebrate genes contain disproportionately large introns compared with exons. Therefore, most random insertions would not directly disrupt coding sequences. In some cases such insertions can be mutagenic by affecting the regulatory sequences, causing mis-expression, or they may result in aberrant splicing or premature termination of transcription.

However, such events often cause hypomorphic mutations, and the null alleles are assumed to occur rarely. Several approaches have been proposed to select for insertions within genes and to enhance the chances of disrupting gene function.

A classic gene trap vector (sometimes referred to as 5' gene trap) harbors a splice acceptor site upstream of a promoterless reporter. When inserted into an intron of a transcribed gene it can cause abnormal fusion transcripts and expression of the reporter gene under regulation of the promoter of the gene into which it is inserted. Because gene trap vectors target introns, they are effective in species with large introns such as vertebrates. However, the rate of alternative splicing through the splice acceptor sequence of the gene trap cassette can be insufficient to prevent fully the synthesis of the original transcript, resulting in generation of hypomorphic alleles or no detectable effect on gene function. Multiple splice acceptor sequences and multiple transcription terminator sequences can be introduced into gene trap vectors to improve their mutagenicity and tagging efficiency.

Another drawback of 5' gene trap is that only insertions in transcriptionally active genes are selected. Genes that are expressed at low levels or are not expressed during the stages observed in the screen remain undetected even if a tagged gene is successfully inactivated by the gene trap. To circumvent this, polyA trap vectors have been designed to tag genes regardless of their transcription status. A basic polyA trap vector (sometimes termed 3' gene trap) generally harbors a promoter driving the transcription of a reporter gene, which has no polyA signal but a downstream 3' splice donor sequence. Such reporters normally produce unstable transcripts, which after successful splicing are polyadenylated using polyA signals of the tagged gene.

A promising modular mutagenesis approach has been recently developed in zebrafish. This strategy combines the mutagenic capacity of 5' gene trap with the high trapping capacity of polyA trap vectors. The 'gene-breaking' transposon vector developed by Sivasubbu and coworkers harbors a 5' mutagenicity cassette and a 3' polyA trap cassette in the same transposon vector. The 5' mutagenicity cassette produces strong termination of transcription by combining an effective splice acceptor sequence with a polyA signal of ocean pout antifreeze protein.

To date, only a few examples of successful gene inactivation using gene trap vectors have been reported in zebrafish. Unbiased systematic studies using a statistically significant number of tagging events are required to estimate the actual mutagenic ability of the various gene trap designs and to compare it with the mutagenicity of transposon insertions carrying simple reporters or enhancer traps.

Transposon vectors can also be designed to screen for gain-of-function mutations by random activation and/or over-expression of genes. In these screens, usually referred to as 'activation tagging' or 'modular mis-expression', transposons are equipped with enhancer or promoter sequences that can induce transcription of the endogenous gene adjacent to the insert. Several laboratories have begun testing these tools in zebrafish, but there are no published reports yet.

MUTAGENESIS OF CLOSELY LINKED GENES BY RE-MOBILIZATION OF GENOMIC TRANSPOSON INSERTIONS

If a transposon insertion residing inside an intron is found to be nonmutagenic, then such a transposon can potentially be re-mobilized in order to produce more insertions in the same gene or to generate flanking deletions. Various transposons preferentially re-transpose to nearby chromosomal sites (for example Ac/Ds in plants, P-element in Drosophila, and Sleeping Beauty in mouse). Several elegant approaches utilize this ability of transposons to inactivate closely linked genes.

Because not all offspring from a founder will harbor re-transpositions, it is beneficial to have a selection scheme for the transposition events. One simple approach is to use a genomic insertion of a gene trap transposon as a donor, and select for gene trap events. A similar selection principle was successfully utilized in the mouse using the Sleeping Beauty (SB) transposon by Keng and coworkers. They used two unique donor sites containing multiple concatemeric copies of the SB element, a feature that ensured high re-transposition rates. The reporter gene carried by the SB element in these donor sites was silenced, and its expression was activated when re-transpositions caused gene trap events. The authors demonstrated high efficiency of this approach for region-specific saturation mutagenesis, producing insertions in all genes within a 4 megabase region surrounding the donor site.

A more effective strategy for selecting re-transposition events uses a combination of two selection markers: one to detect the excision events and another to retain the re-transposed element. This double selection scheme utilizes two types of transposable elements. A 'carrier' element harbors a 'jumper' element and an excision marker. The 'carrier' is used to generate insertions in the genome that are subsequently used as the donor sites for transposition of the 'jumper' element. In plant genetics, such donor sites, usually called 'launch pads', have successfully been used to saturate specific regions of the genome with insertions. Because zebrafish exons are small, the mutagenesis efficiency of this approach is expected to be low. Arming the jumper element with an additional 'mutagenicity cassette' should be beneficial. Another possible approach combines the local re-transposition from a 'launch pad' with a site-specific recombination system such as Cre/lox or FLP/FRT (flipase recombination enzyme/flipase recognition target) in order to produce flanking deletions or inversions. In this method, the carrier and the jumper elements both harbor loxP sites. After the local re-transposition of the jumper element, a fragment between the loxP sites is deleted or reversed according to the relative orientation of the loxP sites. In plants this approach was demonstrated using a combination of the Cre/lox system, T-DNA as the carrier, and Ds element as the jumper. In zebrafish, there are at least three heterologous transposon systems that can be used in various combinations with each other. Tol2 and Ac/Ds can both be used as carrier elements because of their large insert capacity.

Upon mobilization, a transposable element can generate deletions, inversions, and other rearrangements around the donor site. When two insertions of the same transposon are found at different sites on the same chromosome, deletions and inversions of the region between the two insertions can be recovered. Two trans-heterozygous insertions (located at different positions on homologous chromosomes) have also been used to generate deletions in Drosophila. P-element insertions are frequently used as starting-points to generate flanking deletions in Drosophila by screening for male recombination events or 'imprecise excision' events. The former strategy generally relies on the lack of male meiotic recombination in Drosophila. In this approach, recombination between two flanking markers in the male germline occurs only when the P-element undergoes a transposition process called 'hybrid element insertion', which causes frequent deletions. In zebrafish, meiotic recombination in males is also suppressed relative to that of female meiosis, especially near centromeres, but the rate of suppression is unlikely to be sufficient for effective application of this strategy. This approach can potentially be applicable using a pair of closely linked flanking markers. However, it is not known whether similar recombination events occur in fish and what is the frequency of long deletions.

When a transposable element is excised, the double-strand break generated during the excision is sometimes repaired imprecisely, generating flanking deletions or other rearrangements. This has been often utilized in Drosophila. Although, the actual mechanism of 'imprecise excision' is unclear, it should be distinguished from the 'hybrid element insertion', which leaves the transposon at the deletion site. Because large deletions (several kilobases) are expected to occur at low rates, this strategy may be feasible only when the insert is found very close to an exon of the target gene or when there is a convenient selection marker, which can be used to detect deletion events.An improved approach shown in Figure 1f eliminates the need for closely linked markers. A compound element consists of a carrier element that harbors a 'deleter' element inside, flanked by a pair of visual markers (for instance, green fluorescent protein [GFP] and red fluorescent protein [RFP]). During primary screening, the genomic insertions of the compound element are produced using the transposase for the carrier element. If the insert is found close to the gene of interest or if it lands in an intron causing no effect on gene function or hypomorphic mutation, then such insertion can be used as a starting point to produce flanking deletions by activating the deleter element. In this scheme, there is no need for additional genetic markers; deletions on either side of the original insertion can easily be identified by the absence of one of the flanking markers, whereas the other marker shows the presence of the donor site. A similar selection scheme with a single flanking marker was utilized in plants using the Ds transposon as a deleter. The strategies for generating flanking deletions illustrated in Figure 1c,f both have advantages and disadvantages. Until now there have been no reliable data published on the distribution preference of transposable elements in fish. Therefore, to evaluate the feasibility of the approaches described above, it is important to determine the level of preference to insert to linked positions and to determine the frequency and extent of the flanking deletions.

LARGE-SCALE REVERSE GENETICS MUTAGENESIS IN FISH

Saturating the genome with random insertions (obtaining at least one effective hit in every gene) requires generation of a very large number of insertions, significantly larger than the total number of genes in the genome. It is problematic to generate, analyze, maintain, and maximize use of such huge collections.In the classical mutagenesis screens, this problem is addressed in two steps: animals are heavily mutagenized with highly active mutagens such as ENU (N-ethyl-N-nitrosourea) or retroviruses, so that every founder carries multiple mutations; and such collections are screened for the visible phenotypes, and only the mutant strains are retained and analyzed. The main disadvantage of the phenotype-based mutant screening (forward genetics) is that only a relatively small number of genes are targeted, and mutations that do not cause obvious phenotypes are ignored or discarded.

In the post-genome era, the most common problem is gaining insight into the function of genes that have been identified by sequencing but for which mutant information is lacking (reverse genetics). Therefore, reverse genetics approaches to mutant screening must operate with a complete collection of random mutations representing all genes, regardless of the phenotypic manifestation. Until recently, only one reverse genetic approach has been implemented in zebrafish to obtain mutations in the gene of interest. This approach utilizes TILLING (targeting induced local lesions in genomes) and re-sequencing of ENU-mutagenized fish for target-selected screening of point mutations in the gene of interest. Although this approach is very powerful, considerable screening efforts are required each time to obtain mutations in every new gene. Furthermore, this method of mutagenesis generates predominantly missense mutations, which may or may not result in phenotypes.

An alternative approach to reverse genetic mutagenesis screening was recently pioneered by Znomics, Inc. (Portland, OR, USA), who have produced a large-scale library of retroviral insertions in zebrafish. The company has been identifying the locations of retroviral insertions from cryopreserved sperm samples containing, on average, 25 insertions per sample by sequencing the flanking DNA adjacent to the insertions. With 100,000 insertions already identified, Znomics hopes to map a total of half a million insertions. Thus, their bank of retroviral insertions promises to be a valuable resource for the zebrafish research community. However, the actual efficiency of the Znomics screen and the utility of the strains remain to be determined. Despite the recent vector improvements, the frequency of transposon insertions still remains lower than that of retroviral insertions. It is therefore important to determine whether transposon vectors have the potential to complement the retroviral projects in the field of large-scale insertional mutagenesis.

One way to increase the number of transposon insertions per germline is to use the donor lines carrying multiple inserts for re-transposition. The distribution of retroviral insertions (or any other known insertional mutagen) is not absolutely random. Thus, a single insertional mutagen (retrovirus) would be inefficient in hitting genes at 'cold spots', but instead it will produce in vast excess insertions in 'hot spots' when trying to achieve genome-wide saturation. The presence of hot and cold spots narrows the limits for efficient use of a single insertional mutagen in saturation mutagenesis screens. In order to achieve saturation more productively, it is worthwhile to launch several parallel screens of smaller scale using different insertional mutagens, rather than using a single agent alone. In this respect, having an arsenal of several transposable elements is an advantage. Because vertebrate genomes contain only a small proportion of coding sequences, the majority of retroviral insertions (or other random insertions) are found either in the intergenic regions or inside introns. In case of retroviral mutagenesis, the only way to increase the number of exon hits is by significantly increasing the number of insertions.

Unlike retroviral insertions, transposons can be re-mobilized in order to obtain exon insertions or to generate flanking deletions. Re-transposition also offers a unique approach to reducing the size of insertion libraries-the 'launch pad' strategy. A relatively small collection of the launch pads regularly distributed throughout the genome must be generated and deposited in a stock center. Then, even a moderate sized laboratory should be able to use a launch pad that is closest to their gene of interest for region-specific saturation mutagenesis to generate insertions in the target gene.

CONCLUSION

Although thousands of transgenic fish lines can be created, stored, and recorded in a database in individual laboratories, the long-term maintenance of such collections and wide public distribution of the materials by the same laboratories appear problematic. Because large-scale projects in the field of functional genomics promise vast resources for the scientific community, the necessary funding can usually be obtained. In contrast, it is difficult to raise the additional funds necessary to further maintain and distribute the transgenic lines, and therefore the produced materials and data often remain largely unutilized and eventually wasted. Other sensitive issues such as authorship, sharing of intellectual property rights, and oppressive material transfer agreements, among other factors, are common and can significantly hinder progress. Operating on a cost-recovery basis may present a solution, but recovery from the cryopreserved stocks, quality control measures such as prevention of material loss, and insert confirmation can significantly increase running costs. Recently, a private company (Znomics, Inc.) produced a large collection of retroviral insertion strains, which are available for purchase on a per strain basis. However, it is early to speculate whether demand for the transgenic zebrafish lines will be large enough to extend this model to projects of smaller scale.

The distribution problem is unlikely to be resolved without a dedicated stock center that can take on these responsibilities. The success of large-scale functional genomics projects in other model organisms was achieved largely because of centralization. For example, Arabidopsis stock centers maintain hundreds of thousands of genotypes produced by insertional mutagenesis projects, and large-scale collections of transposon insertions in Drosophila were also made available through stock centers.

The existing zebrafish stock center will need to scale-up its capacity in order to accommodate many thousands of new samples that might be generated very soon by several laboratories. At the present time, there appears to be some variability from laboratory to laboratory in cataloging transposon lines. Because various transposon screens are designed for different purposes (mutagenesis, enhancer and gene trapping, driver lines, and so on), it is important to develop a set of standard requirements for the lines that are deposited at the stock center.

7

Transgenic Fish and Their Greater Production

In fish, microinjection is the method most frequently used for gene transfer. However, due to delayed transgene integration this technique almost invariably produces mosaic individuals and if the gene is not integrated into germ cells its transmission to descendants is difficult or impossible. We evaluated the degree of in vivo mosaicism using a strategy where a reporter transgene is co-injected with a transgene of interest so that potential germline founders can be easily identified. Transgenic zebrafish (Danio rerio) were produced using two transgenes, both comprised of the carp-actin promoter driving the expression of either the green fluorescent protein (GFP) reporter gene or the growth hormone cDNA from the marine silverside fish Odonthestes argentinensis. The methodology applied allowed a rapid identification of G0 transgenic fish and also detected which fish were transmitting transgenes to the next generation. This strategy also allowed inferences to be made about genomic transgene integration events in the six lineages produced and allowed the identification of one lineage transmitting both transgenes linked on the same chromosome. These results represent a significant advance in the reduction of the effort invested in producing a stable genetically modified fish lineage.

INTRODUCTION

Transgenesis involves genomic alteration of an organism through insertion, modification or deletion of a gene with the objective of modifying characteristics of interest (Houdebine,

2002; Carter, 2004). In this manner, new stable and genetically determined characteristics can be incorporated into the genome of the receptor organism and possibly transmitted to the next generations. In the last two decades, this technology has been successfully applied in fish due to the fact that these inferior vertebrates present reproductive and biological characteristics that allow easy manipulation of their genetic and physiological processes in the early stages of ontogenesis (Zhu and Shu, 2000). Studies of gene transfer have been carried out in more than 35 teleost species, most of which are important to aquaculture (Zbikowska, 2003). However, genetically modified fish have also been developed as experimental models for biomedical research, especially in studies involving embryogenesis and organogenesis (Amacher, 1999; Motoike et al., 2000; Goldman et al., 2001; Huang et al., 2001; Takechi et al., 2003) as well as in the study of human diseases (Dodd et al., 2000; Dooley and Zon, 2000; Ward and Lieschke, 2002), xenotransplantation (Wright and Pohajdak, 2001; Leventhal et al., 2004; Pohajdak et al., 2004) and recombinant protein production for producing important therapeutic agents (Anderson and Krummen, 2002).

Several techniques are currently available for transgenic fish production which have been developed to increase the efficiency of transgene integration or to produce a large number of transformed individuals simultaneously. Although these new methods of gene transfer are gaining importance due to the encouraging results reported (Tanaka and Kinoshita, 2001; Lu et al., 2002; Grabher et al., 2003; Hostetler et al., 2003; Kinoshita et al., 2003; Kurita et al., 2004), microinjection in fish is still the most successful technique being used due to its simplicity and trustworthiness (Udvadia and Linney, 2003; Zbikowska, 2003). Maclean et al. (2002) stated that microinjection was the best technique to use with tilapia (Oreochromis niloticus). However, when microinjection is used to produce transgenic fish it almost invariably produces mosaic fish due to delayed transgene integration, which occurs only after a few cycles of embryonic cell division. If the transgene is integrated in only one cell group or tissue but not into germ cells the transmission of the gene to descendants is difficult or impossible (Maclean,

1998). The use of reporter genes that allow evaluation of the degree of in vivo mosaicism in transgenic fish can facilitate identification of probable germline founders and the co-injection of a reporter transgene along with the gene construct of interest represents a considerable reduction in the effort needed for the establishment of transgenic germlines. The gene coding for the green fluorescent protein (GFP) from the jellyfish Aequorea victoria has been widely used as a reporter gene because it does not require an exogenous substrate for its activity (Amsterdam et al., 1995) and is stable and non-toxic in receptor organisms (Peters et al., 1995).

The objective of the work described in this paper was to develop an in vivo methodology to evaluate the degree of mosaicism and to identify transgenic zebrafish (Danio rerio) with the potential to generate germlines for the gene of interest.

MATERIAL AND METHODS

Production of Transgenic Fish

Adult wild-type zebrafish (Danio rerio) obtained from a commercial supplier were kept in a closed water circulation system at 28 °C under a 14 h light/10 h dark photoperiod. Freshly fertilized eggs were collected for microinjection and transgenic zebrafish produced using two transgenes containing the carp (Cyprinus carpio)-actin promoter driving the expression of either the A. victoria GFP gene (pc A/GFP plasmid) or the marine silverside fish (Odonthestes argentinensis) growth hormone (msGH) cDNA (pc A/msGH plasmid). The pc A/GFP plasmid was kindly provided by Dr. Suzanne Brooks (University of Southampton, UK) and was used to produce the pc A/ msGH plasmid by replacing the GFP gene with the msGH cDNA (Marins et al., 2002). Both plasmids were linearized with the Spe I restriction enzyme and co-injected at a 1:1 molar ratio into one-cell embryos using a total DNA concentration of 35 ng μL-1.

The linearized transgenes (called c A/GFP and c A/msGH) were transferred in an equimolar ratio to provide the same integration and expression probability due to the fact that they had approximately the same length (c A/GFP = 5.6 kb and c A/msGH = 5.5 kb) and were both under the action of the same promoter.

The microinjection process followed the general protocol recommended by Vielkind (1992) and used an IM-30 motorized picoinjector (Narishige, Japan) to inject approximately 300 pL of DNA solution representing a final number of 106 copies of each transgene per injected embryo. The microinjection needles were produced from Narishige GDC-1 glass capillaries using the Narishige PC-10 micro-electrode puller. A total of 1872 one-cell embryos were injected. Non-injected controls and microinjected embryos were incubated at 28 °C until hatching.

Evaluation of mosaicism in the first transgenic generation (G0) One week after fertilization the fish larvae were analyzed by epifluorescence microscope (excitation = 485 nm; emission = 520 nm) and classified according to their GFP expression patterns (Gibbs and Schmale, 2000; Thermes et al., 2002) as follows: weak = few cells expressing GFP; moderate = less than 50% of the body expressing GFP; or strong = more than 50% of the body expressing GFP.

Evaluation of the msGH Gene Expression by RT-PCR

To confirm G0 msGH gene expression we used the reverse transcriptase polymerase chain reaction (RT-PCR) to analyze several four-week old fish expressing GFP. Total RNA was extracted by humanely sacrificed 12 G0 fish and homogenizing their bodies in TRIzol reagent (Invitrogen, Brazil) according to the protocol suggested by the manufacturer. Approximately 3 μg of total RNA from each fish was used as template for the RT-PCR with the AP primer (5'-GGCCACGCGT-CGACTAGTAC(T)17-3', Invitrogen, Brazil). The complementary DNA (cDNA) synthesis was carried out using the enzyme RT SuperScript III (Invitrogen, Brazil) according to the protocol suggested by the manufacturer. The cDNA obtained was used as template for the msGH gene amplification using the specific primers EXO 293 (5'-GAAAGCTCTCTGCAGACGGA G-3') and GHEX6-RIG (5'-AGAGTGCAGTTTGCCTCT GG-3'), which produce a 467 bp msGH fragment but do not amplify the endogenous zebrafish growth hormone gene. PCR was carried out in a 12.5 μL reaction volume containing 1.25 μL of 10X PCR buffer, 0.2 μM of each primer, 0.2 μM of each dNTP, 0.75 mM of MgCl2, 0.5 unit of Platinum Taq DNA polymerase (Invitrogen, Brazil) and 1 μL of cDNA solution. The reaction was incubated at 94 °C for 2 min followed by 30 cycles of 15 s at 94 °C, 30 s at 65 °C and 30 s at 72 °C, and a final elongation step of 5 min at 72 °C and the PCR products separated on 1% (w/v) agarose gel stained with ethidium bromide (0.5 μg/mL) and visualized by ultraviolet transillumination.

Assessment of the Biological Effect of msGH in the G0 Generation

The objective of this experiment was to verify whether the msGH transgene produced significant biological effects in the growth performance of the transgenic individuals. Groups of non-transgenic and G0 transgenic fish classified as strongly expressing GFP were reared until one-year old for comparison of their final average weight. Fish were fed ad libitum twice a day with a commercial fish food (ColorBits, Tetra) containing 47.5% protein. The average weight data were statistically analyzed using a t-test for heterogeneous variances contained in the Statistica program v. 6.0 (Statsoft, USA).

Transmission of Transgenes to the Second (G1) and Third (G2) Transgenic Generations

A group of larvae showing strong GFP expression were selected and reared until sexual maturity as being probable germline founders. For the transgene transmission study, eight G0 mosaic fish classified as strong for GFP expression were separated in individual aquariums and crossed with non-transgenic fish. Two of these transgenic fish did not developed reproductive behavior. Six sexually mature transgenic fish were mated with wild-type fish to produce G1 offspring which were assessed by epifluorescence microscopy as described above to verify the presence of the uniform GFP expression expected after genomic integration. To confirm the presence of the msGH transgene in the G1 fish genome, GFP positive larvae were cultivated until adulthood and a small fin clip was taken from each fish and the genomic DNA extracted (Sambrook et al., 1989). The msGH gene was amplified using the specific primers

EXO-293 and GHEX6-RIG and the PCR conditions described above. Only GFP positive G1 fish were tested for msGH because the objective was to identify G0 fish which were transmitting both transgenes to the descendants.

Twelve G1 fish, six from each of two G1 parents carrying both transgenes, were crossed with wild-type fish to produce G2 progeny, which were assessed using epifluorescence microscopy as described above. To verify whether or not the transgenes had integrated on the same chromosome, sub-samples of GFP positive and negative fish (N = 12) were randomly selected from each lineage and their DNA extracted and subjected to PCR, using the conditions described above, to detect the presence of msGH.

RESULTS

Transgenic zebrafish were produced by the co-injection of 1872 one-cell embryos, using the transgenes c A/GFP and c A/msGH in an equimolar ratio (1:1). At the time of assessment by epifluorescence microscopy, one week after fertilization, the survival rate for untreated fish embryos was 1414 out of 1872 (75.5%) while the survival rate of the microinjected embryos was 877 out of 1872 (46.8%), of which 275 out of 877 (31.4%) were classified as GFP negative (no expression), 315 out of 877 (35.9%) as weakly GFP positive, 198 out of 877 (22.6%) as moderately GFP positive and 89 out of 877 (10.1%) as strongly GFP positive. The sum of the three GFP expression classes was 602 out of 877 fish (68.6%).

All of the 12 one-month old G0 GFP positive fish assessed for msGH expression by RT-PCR were msGH transgene positive. The analysis of the average weight after 12 months demonstrated a significant increase ($p < 0.01$) in the transgenic group (1.79 g ± 0.37) in relation to wild-type fish of the same age (0.68 g ± 0.13). This represents an increase in growth of 2.6 times for the transgenic group as compared with the non-transgenic fish, and shows the biological effect of expression of the c A/msGH transgene in zebrafish.

In the transgene transmission study, eight G0 mosaic fish classified as strongly expressing GFP were crossed with non-transgenic fish. Two of the transgenic fish did not developed reproductive behavior but two males (M0104 and M0204) and four females (F0104, F0204, F0304 and F0404) reproduced, four of which (M0104, F0104, F0204 and F0304) transmitted the GFP gene to the G1 in percentages that varied from 2.2% to 42%. The GFP expression pattern observed in the G1 offspring showed fish expressing the transferred gene in all body cells. However, the msGH gene was detected only in G1 descents obtained from the M0104 and F0104 parent fish. Only half the GFP-positive offspring of the M0104 parent were also carrying the msGH gene but for the F0104 parent all the GFP-positive offspring were also positive for the exogenous GH gene. No GFP gene expression was detected in the offspring of the M0204 and F0404 parents.

The G1 offspring of the M0104 and F0104 G0 mosaic parents which were positive for both transgenes were reared until sexual maturation and six G1 fish from each parent were mated with wild-type fish to produce G2 offspring. A total of 466 G2 offspring resulted from the M0104 matings while 588 G2 offspring were obtained from the F0104 matings. These results are summarized in Table 1 and indicate two different transmission patterns for each lineage. For the male M0104 parent the G2 offspring consisted of 25% negative for both transgenes, 25% GFP-positive only, 25% msGH-positive only and 25% GFP/ msGH-positive, while for the female F0104 parent the G2 offspring consisted of 50% negative for both transgenes and positive for both transgenes.

DISCUSSION

Although some methodologies have presented promising results in increasing first generation (G0) transgenic fish production, there is still the problem of rearing large numbers of G0 fish and the subsequent identification of those with the potential to generate stable germlines carrying the active transgene. This process is even more difficult for large fish such as carp, salmon, tilapia or trout which need very complex culture facilities. The methodology applied in our study not only allowed the production of transgenic fish carrying the active transgene of interest (c A/msGH) but also the reporter transgene (c A/GFP) which allowed the evaluation of mosaicism in all the G0 transgenic fish generated. The analysis of the GFP expression patterns permitted the selection of possible germline founders from the fairly low number of fish in the G0 generation.

A week after microinjection 68.6% of the fish embryos expressed GFP, which represents a high efficiency of transgenic fish production. However, part of this observed expression can be attributed to transitory expression due to the transcription of unintegrated transgenes (Chong and Vielkind, 1989). Our results are significant when compared to the 1.95% transgenic zebrafish obtained by Morales et al. (2001) and the 10% transgenic tilapia obtained by Rahman et al. (1997) using the same reporter transgene co-injection strategy. We found that 10% of the fish analyzed by us presented strong GFP reporter gene expression, significantly more than the 5% with strong GFP expression reported by Gibbs and Schmale (2000) for G0 transgenic zebrafish and the 3% with strong GFP expression reported by Thermes et al. (2002) for G0 transgenic medaka. The conditions used by these authors were similar to ours and they also used linearized transgenes in which the GFP gene was controlled by ubiquitous promoters (and-actin).

Although our RT-PCR analyses showed that 100% of the G0 GFP positive fish were expressing the msGH gene not all these fish were carrying the msGH transgene in their germ cells and could transmit the msGH transgene to the next generation, this being evident when the G0 and wild-type fish were crossed. The growth experiment data supported our RT-PCR results and indicated that the transgenic group increased in weight 2.6 times more than the non-transgenic group. These results demonstrate that the c A/msGH transgene was producing an active hormone. The higher weight of the transgenic fish was probably related to increased circulating msGH which could not be controlled by the negative feedback mechanism which regulates endogenous GH gene expression (Peter and Marchant, 1995).

The main negative consequence of mosaicism in transgenic fish germline production is the fact that the germ cells of G0 fish can receive few or no copies of the transgene, making transgene transmission to the next generation difficult or impossible (Maclean, 1998). However, this problem can be minimized by evaluating mosaicism using a reporter transgene co-injection strategy. This is supported in our study by the presence of strongly GFP-positive fish (2 out of 6, or 33.3%) transmitting both transgenes to the G1 generation, indicating a considerable increase in the possibility of identification of germline founder fish. According to Maclean (1998), the rate of transgene transmission to the G1 is low, with only approximately 5% of G0 transgenic fish having the capacity to transmit transgenes to the next generation.

The percentage of GFP positive G1 fish produced in our study indicated the degree of mosaicism in the germ cells of the G0 fish. We found that two fish did not transmit any transgene to their descents, indicating that the transgenes were not integrated into the germ cells, while four fish transmitted the c A/GFP transgene to produce 2.2% to 42% of GFP-positive descents. Therefore, despite the fact that some of our fish strongly expressed GFP, transgene integration into germ cells was extremely variable. In theory, if all the germ cells of a transgenic fish contained the transgene (i.e. no germ cell mosaicism) and this fish was mated with a wild-type fish 50% of the offspring would express the transgene.

Maclean (1998) pointed out that generally only a small percentage of the offspring from G1 mosaics are transgenic, which makes identification difficult when an easy detectable marker is absent. In our study, the identification of G1 transgenic fish was greatly facilitated by the presence of the GFP gene reporter and simple evaluation using epifluorescence microscopy allowed rapid identification of the transgenic fish. Additionally, G1 fish originating from the F0104 female and the M0104 male (which were carrying both transgenes) were crossed with wild-type fish to verify how the transgenes integrated in the G2 fish genome. In the G2 produced from the M0104 lineage a number of GFP-positive fish did not carry the exogenous GH transgene while some GFP-negative fish were carrying it. This indicates that the c A/GFP and c A/msGH transgenes segregated in the G2, since the observed genotypic ratio is in accordance with genes situated on different chromosomes. However, for the F0104 lineage G2 descents all the GFP-positive individuals were carrying the c A/msGH transgene as well as the c A/GFP transgene, indicating that both transgenes had been integrated on the same chromosome.

The methodology described in this paper allowed the rapid identification of G0 transgenic fish and also identified which fish were transmitting transgenes to the next generation. This strategy also allowed inferences to be made regarding genomic transgene integration events, and permitted the identification of the one lineage (out of the six produced) which contained and transmitted both transgenes linked on the same chromosome. These results represent a significant advance in the reduction of the effort involved in the production of genetically stable modified fish lineages.

8

Risks and Challenges

For centuries farmers have selectively bred crops and livestock to produce the desirable phenotype and increase yields. Advances in the study of genetics have enabled us to modify actual genes on a strand of DNA to achieve the desirable traits. Despite the benefits, genetically modified organism (GMO) technology involves ethical, religious, cultural, social, and most importantly, environmental risks. The introduction of genetically modified fish may have great ecological impact on natural environments, for example, they could introduce diseases, or threaten the survival of their conspecifics by competing for limited resources. The fact is, we do not know all the ecological impacts from the introduction of transgenic fish. Unless we know all the effects, we must accept the concept of precautionary principles to avoid serious and irreversible damage to the environment.

INTRODUCTION

Advances in the study of genetics or advances in the understanding of the structures and expression of genes has enabled us to develop artificial genes programmed for specific purposes, which may be transferred into animal or plant cells. Animals and plants developed from these cells and their offspring containing these artificial genes are referred to as transgenic (Sin 1997). Transgenic organisms are also referred to as genetically modified organisms (GMOs), genetically engineered organisms (GEOs), and genetically altered organisms (GAOs). Kapuscinski and Hallerman (1990) define transgenic fish "as fish bearing-within their chromosomal DNA-copies of novel genetic constricts introduced through molecular genetics techniques."

For centuries farmers have selectively bred plants and animals to achieve the most desired phenotype and to increase yields. Gregor Mendal, an Austrian monk of the mid-1800s, experimented with and approached selective breeding in a more scientific manner. Mendal's experiments with garden peas initiated the science of genetics (Anderson 2001). In the 1900s the development of hybrids were becoming increasingly important commercially. Today, geneticist can modify actual genes on a strand of DNA to achieve desirable traits. Most of the plant and animal species that people rely on for their food supply have been genetically selected in order to improve their performance and quality. By late 1999, 60 percent of the processed foods in the United States were derived from genetically modified foods (Pollack and Shaffer 2000). Herbicide-resistant crops and virus resisting crops have been developed by researchers to improve the agriculture industry (Anonymous 2001). Researchers also predict the production of genetically engineered livestock will offer improved nutritional value and the production of safer animal products for human consumption through strategies that might increase disease resistance (Anonymous 2001). This improvement in biotechnology offers the same results as did the traditional breeding methods but in a more precise and efficient manner.

DISCUSSION

Transgenic fish are produced by the artificial transfer of rearranged genes into newly fertilized eggs using the preferred method of microinjection. This process of microinjection first calls for the construction of artificial genes or fusion genes built for a specific genotype (Sin 1997). These genetic constructs are composed of genetic rudiments derived from genes of the same species or unrelated species. This newly constructed DNA is microinjected into a host chromosome before first cell division so that the desired gene would be present in every cell. An alternative method is the microinjection of the gene complex at some developmental stage after the first cell division. This results in the development of an organism containing the desired gene in only a subset of their cells (Kapuscinski and Hallerman 1990).

Transgenesis is a fast way of mass producing animals with new and desirable genetic traits, and is therefore, a desirable method for aquaculture and marine management. "Using microinjection as a gene transfer method has produced transgenic fish with enhanced growth rates in common carp (Zhang et al. 1990; Chen et al. 1993), northern pike (Gross et al. 1992), Atlantic salmon (Du et al. 1992), coho salmon, rainbow trout and cutthroat trout" (Delvin et al. 1994, 1995). These transgenic fish that reach adult size in a much shorter period could have obvious appeal with the Canadian aquaculture industry, which earns $600 million a year from the sale of farmed fish (Nichols 2001). The development of this "supersalmon" would mean more fish for more people at a lower price due to the fact that fish farmers could double production without doubling costs (Golden and Thompson 2000). Current researchers are also concentrating on producing a freeze-resistant salmon by using a gene construct from an Arctic fish species known to resist freezing by producing an antifreeze protein in the blood (Davies et al. 1989).

Fletcher et al. (1988) used an antifreeze gene construct from the winter flounder. This may allow farming of these salmon in cold-water environments. "The Canadian Food Inspection Agency (CFIA) predicts that in the next ten years they could develop new gene constructs that could "cause overexpression of hormones such as prolactin that are involved in the transformation of anadromous fish from salt to fresh water, thereby making it theoretically possible to raise marine fish in fresh water." They may also be able to manipulate the length of reproductive cycles, "expand the tolerance of aquaculture fish to a wider range of environmental conditions, improve resistance to a variety of pathogens," control sexual maturation, sex differentiation and sterility (Anonymous 2001). Despite all the benefits, GMO technology involves ethical, religious, cultural, social, and most importantly, environmental risks. Our natural food sources have evolved over millions of years to best adapt to our natural world therefore, raising the ethical issue of whether we should be altering nature. Also, many religions believe that the natural world is sacred and that we, as humans, should not artificially alter it (Anderson 2001). Due to these concerns, businesses that produce GMOs for the public risk refusal to buy GMO products by consumers.

The greatest concern is the risks posed by the release or escape of genetically modified organisms on the environment. Despite all the benefits of the production of genetically modified fish, their introduction may have great ecological impact on natural environments. Farmed fish tend to escape the fish farm pens into rivers and oceans where they have the ability to compete in the wild (Nichols 2001). "Any release of transgenic organisms into nature is a concern because ecological relationships between genetically engineered organisms and other organisms (including their wild-type conspecifics) are unknown" (Muir et al. 2001). Released fish or fish that have escaped could introduce new diseases or compete for limited resources. Also, if a sterile transgenic male fish mates with a native female, her reproductive effort is wasted (Muir et al. 2001).

Transgenic fish may threaten the survival of their conspecifics as well as other species in their community due to some new advantage. For example, the salmon genetically engineered to grow ten times the rate of normal salmon might compete so vigorously in the wild that they could kill off natural salmon stocks, by the advantage of predator avoidance or by the ability to outcompete their conspecifics for food and nutrients. Devlin et al. (1999) performed a study on the feeding habits of the growth hormone-transgenic coho salmon, concluding that the growth hormone increases feeding motivation and appetite giving them the ability to outcompete native salmon for food. Also, because these transgenic salmon reach reproductive stage earlier than natural salmon, they could have the ability to soon outnumber natural stocks (Nichols 2001). Along with these direct effects that may occur from the release of transgenic fish, many ecological effects may be indirect taking some time to appear. Therefore, the absence of immediate effects does not ensure that no effect will occur.

"With both benefits and risks, we are working in the realm of the potential and the unknown. Huge uncertainty and unpredictability are present, which of course creates risks within themselves" (Anderson 2001). The biggest unknown is the effects from the release of genetically engineered fish into the ecosystem. These effects could be uncontrollable, permanent, and irreversible, and by the time they are present it will be too late. The assessment of the benefits and risks associated with the production of genetically engineered fish requires expertise from many different fields of study including molecular biology, cell biology, evolutionary biology, genetics, physiology, population and community ecology, ecosystem science, and more (Hodgson and Sugden 1988).

The survival and reproduction of transgenic fish, their interaction with native fish, and their impact on the environment will all need to be measured and addressed before anyone can confidently identify all risks (Tiedje et al. 1989). This is virtually impossible to do in one study, because different genetically modified fish will have different affects, and the same species could have different affects in different habitats. The complexity of the impacts between the environment and genetically modified fish may include indirect effects and time lags therefore, it is difficult to predict all impacts (Kapuscinski and Hallerman 1990). All models that have been used in evaluating the potential effects of the introduction of transgenic species into the environment have some merit, but all have limits in their applicability (Tiedje et al. 1989). Also, studies performed in laboratories are not completely accurate because the complexity of the physical, chemical, and biological environment cannot be duplicated (Tiedje et al. 1989).

Since all effects of the introduction of transgenic fish are unknown, proper risk management practices need to be applied. The coordination of international risk assessment is necessary due to the fact that ecological effects and geographical ranges of organisms exceeds international boundaries (Tiedje et al. 1989). Lofstedt et al. (2002) explain the concept of precautionary principles and how they "have emerged as a fundamental challenge to conventional policy analysis, for risk-related activities surrounded by great uncertainty." Germany and Sweden had early acceptance of the concept of precautionary principles, for example, the 1969 Swedish Environmental Protection Act required industry to demonstrate the safety of environmentally hazardous activities. Precautionary principles offer hope of avoiding serious and irreversible damage to the environment, but for some, precautionary principles is an infringement on science. The validity of precautionary principles has deteriorated as science has advanced (Lofstedt et al. 2002). I believe that GMO technology should not advance without the application of precautionary principles. All risks must be assessed before the introduction of transgenic organisms, specifically transgenic fish. If transgenic fish are going to be produced and raised, the process must be done with minimal risks.

Kapuscinski and Hallerman (1990) rated in order of increasing risk some recipient environment possibilities: indoor research laboratories; above-ground tanks or raceways at contained outdoor research facilities; levee or dug-out ponds at research facilities; commercial-scale operations in outdoor tanks or raceways; commercial operations in levee or dug-out ponds; research or commercial cages and net pens in natural aquatic ecosystems; and natural aquatic ecosystems. Risks could be lowered to almost zero if the transgenic fish are raised in closed tanks.

CONCLUSION

Genetically modified fish pose economic benefits to aquaculture, but also pose many ecological risks. Since most of the risks are unknown, further research, review, and monitoring is important. Conducting thorough studies under near natural environmental conditions provides our only hope of assessing the risks associated with the introduction of genetically modified fish. Demand to produce transgenic fish is only increasing, so to improve the chances that decisions to use transgenic fish outside the laboratory are made in an ecologically sound manner, ecological assessments and studies need to be initiated at this time. Minimization of the risks associated with genetically modified organisms can be addressed through the processes of risk assessment and risk management, and by the acceptance of precautionary principles. If we do not approach this subject with caution, the introduction of genetically modified fish could have detrimental affects that are uncontrollable, permanent and irreversible.

In the field of transgenics, there have been numerous methods employed in the transfer of foreign DNA to target organisms, and each has had varying success depending upon the recipient animal. To date, microinjection is still the preferred method for aquatic species, and many of the techniques are modified versions of those used in mammalian microinjection. However, significant differences in fish egg physiology (i.e., a large yolk, an impenetrable outer membrane, and overall opaqueness) make it impossible to inject DNA directly into the nucleus, which renders this method time consuming, labor intensive, and technically challenging.

Electroporation is an alternative that alleviates some of these difficulties and could greatly facilitate research and commercial development of transgenic animals. However, techniques for electroporation have not been standardized, and considerable variation exists among methods. Initial studies involving electroporation in aquatic species used a direct current (DC) with either a pulse of exponential decay or multiple rectangular pulses; however, levels of transfer were low, and expression questionable. As a modification to cell culture electrotransfection techniques, Chang combined the properties of a DC field and an oscillating field to produce a new waveform, called DC-shifted radio frequency (RF) pulses. The idea is that this new waveform would be gentler to the membrane and therefore increase the probability of the membrane resealing. We recently tested the efficacy of DC-shifted RF pulses to improve gene transfer in fish embryos and to determine if transgene expression could be achieved. The Japanese medaka (Oryzias latipes) was chosen as the study organism. Medaka offer several advantages as model organisms. For example, the medaka is a small, egg-laying freshwater teleost with a rapid generation time, large number of offspring, and external fertilization. Many aspects of this species have been well characterized, including physiology, genetics, and development. Medaka are easy to maintain in a laboratory setting, fertilization time can be predicted, and the fertilized embryos develop externally.

We used the commercially available Gene Pulser II and RF module (Bio-Rad Laboratories, Hercules, CA) with the reporter gene constructs pCMV-SPORT-ß-gal (Invitrogen, Carlsbad, CA) and the pEGFP-1 promoterless plasmid linked to a gene of agricultural importance. The pCMV-SPORT-ß-gal plasmid contains the ß-galactosidase gene from E. coli driven by the CMV promoter and followed by the SV40 polyadenylation signal. The EGFP construct consisted of the CMV promoter driving expression of both the egfp gene and the phyA phytase gene from Aspergillus niger. Plasmid DNA was purified using a commercial kit. Following fertilization, embryos were collected as quickly as possible. Embryos were rinsed with deionized water, separated into groups, and placed into a 0.4cm gap-width cuvette with supercoiled plasmid DNA resuspended in buffer.

Over 15,000 embryos were electroporated with pCMV-SPORT-ß-gal in order to determine how each parameter (buffer choice, buffer volume, quantity of DNA, total voltage, percent modulation of the wave, the radio frequency of the wave, number of bursts, length of bursts, time between bursts, and number of embryos) affected gene transfer. Following electroporation, embryos were rinsed with deionized water, placed within a homemade hatchery, and allowed to develop. At two weeks of age, fry were sacrificed, fixed, and stained overnight to detect ß-galactosidase activity. Transgenic fry were compared to wild type controls, and blue tissue staining was used as an indicator of transgene expression. Total voltage significantly affected survival of electroporated embryos, and this effect was influenced by buffer choice and volume. The radio frequency significantly affected the number of transgenics obtained, and this effect was most significant with 100% modulation of the wave. The length of the bursts as well as the amount of time between bursts also significantly affected the number of transgenic fry obtained.

Based upon these results, we concentrated on the effects of voltage, radio frequency, burst duration, and burst interval on gene transfer in embryos electroporated in groups of 50 with 4mg of plasmid DNA resuspended in 400ml of HEPES buffered saline solution. From these results, it is evident that voltage will not only affect survival, but in turn will also affect the probability of surviving fry being transgenic. For the most part, survival decreased as voltage increased, yet the percentage of fish with transgene expression increased as voltage increased up to 20-25V and then decreased. Similarly, as burst duration increased, survival decreased. However, the effect of burst duration on the production of transgenic fish was dependent upon the radio frequency. When the radio frequency was between 35 and 45KHz, the highest percentage of transgenics was obtained with a burst duration of 10msec; for a radio frequency of 30 or 50KHz, the highest percentage of transgenics occurred with a burst duration of 50msec. In contrast, the effect of the burst interval on the production of transgenic fish was linear and did not appear to be influenced by any other factor. As the burst interval increased, the efficiency of gene transfer decreased.

Since the expression patterns in the ß-galactosidase stained fish displayed a high level of mosaicism, we decided to test for germ-line integration, transmission, and expression of the transgene in subsequent generations. The EGFP construct was utilized because its expression can be observed in living organisms with minimal discomfort to the fish. Following analysis of the above parameters and their effects, we decided to use the same DNA concentration resuspended as above, 25V, 100% wave modulation with a radio frequency of 35KHz, and 3 bursts for 10msec each with a 1.0sec pause between bursts. These fish were grown to sexual maturity and a double back-cross performed. Approximately 85% of the founders that survived to sexual maturity demonstrated germ-line transmission to their offspring3. The observance of green fluorescent protein indicated transgene expression, and Southern blots demonstrated heritability of both the copy number and junction fragments of the transgene in a Mendelian fashion.

It is evident that electroporation has the potential to increase greatly the efficacy of transgenic fish production. We were able to use an electroporation device based upon the principle of a direct current with an oscillating field, called DC-shifted RF pulses, to produce effectively and efficiently germ-line transgenic medaka. Unlike microinjection, this technique is very rapid and easy to use, allowing for as many embryos as can be collected to be electro-porated in a day. Although some time is required to optimize the electroporation condition, once a set of conditions has been determined, it can be used repeatedly to produce transgenic fish with varying transgenes. The high percentage of transgenic fish obtained should allow for an abundant number of founders to be screened for proper or ideal expression, as well as to decrease the number of embryos needing to be sacrificed in order to obtain one transgenic one. Unfortunately, the RF module from Bio-Rad Laboratories is no longer commercially available.

9

Cryopreservation of Marine Fish Sperm

Since the first work of Blaxter in 1953, fish sperm cryopreservation has been attempted on about 30 marine species. The present paper reviews the techniques used and the results published in these species. Particular attention is paid to the handling procedure of sperm before freezing, the problems of semen ageing and semen contamination with urine. The quality of frozen-thawed semen was evaluated using previously standardized biotests, such as a two-step motility activation technique adapted for the different species and fertilization assays using a discriminating insemination technique. Most extenders used in marine fish are saline or sugar solutions. From the investigated cryoprotectants, dimethyl sulphoxide (DMSO) generally leads to the best results. Cooling rates range from 8 °C to 99 °C min-1; the thawing rate is generally high. Compared with freshwater species, a high percentage of spermatozoa survives cryopreservation. Therefore, and because of the simplicity of the techniques, the cryopreservation of marine fish sperm is suited for application in aquaculture.

INTRODUCTION

Techniques of sperm management have been established in some freshwater fish species such as cyprinids (Billard, Cosson, Crim & Suquet 1995) or siluroids (Legendre, Linhart & Billard 1996) and in salmonids (Scott & Baynes 1980; Billard 1992). Among these techniques, sperm storage and cryopreservation are of special interest. At 0°C conditions, spermatozoa can be stored for a few hours up to several days, depending on the species while cryopreserved gametes can be theoretically stored between 200 and 32 000 years without deleterious effect (Ashwood-Smith 1980).

The use of cryopreserved spermatozoa can be delayed from the date of collection and adjusted to the moment of ova processing. The benefits of this technique include:- Synchronization of gamete availability of both sexes: ovulations are only noticed when sperm production declines in cross fertilization of different strains like spring and autumn spawning herring (Clupea harengus L.: Blaxter 1953).

Use of the total volume of available semen: this is useful for sperm economy in species where semen is difficult to obtain (Japanese eel, Anguilla japonica Temminck & Schlegel) (Ohta & Izawa 1996), but also in species where only low volume of semen can be stripped in captivity (yellowtail flounder, Pleuronectes ferrugineus L.: Clearwater & Crim 1995 or turbot, Psetta maxima L.: Suquet, Billard, Cosson, Dorange, Chauvaud, Mugnier & Fauvel 1994).-Simplifying broodstock maintenance: off season spawning can be induced in most cultured fish species, by the manipulation of photoperiod and temperature cycles (Bromage 1995). However, the technique is cost intensive. When cryopreserved sperm is available all year round, the manipulation of the spawning season could be restricted to females.-Transport of gametes: useful when male and female gametes are collected in different locations. This enables also the introduction of genes from the wild into hatchery stocks.

Avoiding aging of sperm: the senescence of sperm during the course of the spawning season has been reported for many fish species and results in a decrease of milt quality (Rana 1995a).

Cryopreservation allows the collection of sperm when it has the highest quality.

Experimental programs: for genetic studies, in comparing the breeding performances of successive generations in the same experiment and for experiments where the use of identical sperm samples is necessary over an extended period e.g study of short term storage of ova. Conserving genetic variability in domesticated populations: the use of a limited number of breeders leads to a reduction of heterozygosity. The cryopreserved semen of selected strains or genetically improved populations can be introduced in domesticated stocks e.g. the sperm of sex reversed gynogenetic hirame (Paralichthys olivaceus Temminck & Schlegel) female (Tabata & Mizuta 1997). Gene banks of cryopreserved semen can also be used to maintain genetic diversity of fish populations that are endangered and protect against inbreeding. In protogynous hermaphrodite species such as Black grouper (Epinephelus malabaricus Bloch & Schneider), sperm can only be collected in 5 to 10 years old animals (Gwo 1993). As a consequence, success in breeding is greatly enhanced by the use of frozen sperm.

Under consideration of the many benefits of this technique, the sperm of over 200 fish species has been cryopreserved (for review see Billard et al. 1995). These techniques have been reviewed in detail by Stein & Bayrle (1978), Scott & Baynes (1980), Stoss (1983), Chao (1991), Jamieson (1991), Tiersch (1995), Rana (1995a & b), Maisse (1996) and Maisse, Labbé, Ogier de Baulny, Leveroni & Haffray (1998). Most publications are devoted to salmonids, tilapias and carps. In marine fish, the first report was by Blaxter (1953) in herring. As marine fish farming expands, there is an increasing need to apply sperm cryopreservation techniques. Due to the increasing knowledge in sperm biology also spermatozoa of other marine fish have been recently cryopreserved successfully. This paper reviews the techniques and results obtained for cryopreservation of marine fish species. It emphasizes the high survival and fertilization capacity generally obtained in frozen-thawed spermatozoa of marine fish in comparison to freshwater species. Special attention was paid on individual variations of sperm quality and on biotests useful to determine the quality of thawed cells.

INDIVIDUAL VARIATIONS OF SPERM QUALITY

High individual variations of milt quality are frequently reported (Rana 1995a). This may be due to genetic variability but also to sampling location (from testes to the genital pore), contamination of sperm samples by urine and aging of spermatozoa during the spawning season. Similar to the Salmonidae (Morisawa & Morisawa 1988), a maturation process of spermatozoa has been reported for several marine fish species. Compared to samples collected by stripping, the percentage of motile spermatozoa of turbot decreases when intratesticular milt is collected. In Japanese eel, testicular spermatozoa exhibited only a very slight motility after activation, while spermatozoa of stripped sperm could be successfully activated (Ohta, Ikeda & Izawa 1997a). Furthermore, testicular spermatozoa of this species, incubated in saline solutions with high concentrations of K_+ and HCO_3^-, artificially acquire their motility potential and show a similar fertilization capacity as stripped milt (Ohta, Tanaka, Kagawa, Okuzawa & Iinuma 1997b).

Because of the close vicinity of sperm ducts and urinary ducts, sperm samples are frequently contaminated with urine. This phenomenon is described in freshwater fish species (Rana 1995a), but more scarcely in marine species. In turbot, a mean contamination rate of 15.3% (urine volume: sperm volume) was reported, basing on the determination of urea concentration (Dreanno, Suquet, Desbruyères, Cosson, Le Delliou & Billard 1998). Artificial urine contamination of milt decreased the percentage of motile spermatozoa, sperm velocity, the fertilization capacity and the storage ability. These effects increased with urine concentration and incubation time. Catherization of the ureter prior to sperm collection significantly decreased the urine contamination to 9.3%. In species with high sperm densities such as sea bass (Dicentrarchus labrax L.), urine contamination is more easily detected because of the lower viscosity and colour changes of portions of sperm samples containing urine (Fauvel, Savoye, Dreanno, Cosson & Suquet 1999).

Intratesticular aging of sperm has been reported for many fish species and it affected sperm quality at the end of the milting period (Rana 1995a). In marine fish, in the sea bass the concentration of spermatozoa decreased as the spawning season progressed (Fauvel et al. 1999). Spermatozoa of this species maintained their swimming duration for a longer period at the beginning of the milting season than at the end (Billard, Dupont & Barnabé 1977; Sorbera, Mylonas, Zanuy, Carillo & Zohar 1996). Furthermore, lower motility rates, fertilization rates and reduced short-term storage capacity were recorded at the end of the reproduction period. As revealed by electron microscopy, also the structure was changed in aged sea bass sperm (Dreanno, Suquet, Fauvel, Le Coz, Dorange, Quemener & Billard in press). In the Atlantic halibut (Hippoglossus hippoglossus L.), sperm motility was reduced at the end of the reproductive period (Methven & Crim 1991; Shangguan 1998). The highest motility rates of winter flounder (Pleuronectes americanus Walbaum) sperm were found at the beginning of the milting period (Shangguan & Crim, 1995). A decline of motility parameters, storage and fertilization capacities as well as alterations of the plasma membrane were reported in turbot as the milting season progressed (Suquet, Dreanno, Dorange, Normant, Quemener, Gaignon & Billard 1998a). As recorded in sea bass and turbot, senescence of spermatozoa can result in a decrease of freezing ability.

Subsequently, when sampling sperm for cryopreservation, one must take care of the three described factors (intra testicular maturation, contamination by urine and aging of sperm) as they can decrease the initial quality of spermatozoa.

BIOTESTS

Several biological tests are used to improve the cryopreservation techniques and to assess the sperm viability of frozen-thawed spermatozoa. An efficient quality test for frozen-thawed spermatozoa must be correlated with the fertilization capacity. The test must allow objective and sensitive measurements of a large number of cells and give a rapid answer because of the fast deterioration of ova quality immediately after collection. The presently applied tests describe physiological parameters (percentage of motile spermatozoa and fertilization capacity). Complementary biochemical tests (ATP, O_2 consumption) or morphological ones (electron microscopy) precisely localize damages due to the cryopreservation process.

Motility Assessment

Billard & Cosson (1992) reviewed and improved early techniques used to describe fish sperm motility: because of a high sperm concentration in milt, the assessment of sperm motility requires a high dilution rate. A two-step procedure is necessary to initiate simultaneous motility of the maximum number of spermatozoa: sperm is first diluted in a medium that does not initiate motility. Then, sperm movement is activated, by mixing spermatozoa directly under the microscope with an activating solution.

Cosson, Billard, Cibert, Dreanno, Linhart & Suquet (1997) developped the techniques allowing a fine description of motility. This observation is greatly enhanced using dark field optics microscopy attached to a camcorder. Spermatozoa can be visualised on a video monitor. Using stroboscopic illumination, the flagellar beat frequency can be measured by reference to the calibrated frequency of the flash illuminator. The percentage of motile cells is estimated using an arbitrary scale, each step representing a wide range of spermatozoa showing progressive movement. CASA (computer-assisted sperm analysis) system allows an analysis of videotapes using software which set up parameters are adjusted for the studied species sperm characteristics. Straight or curvilinear velocities are rarely assessed using CASA system (turbot, Dreanno, Suquet, Quemener, Cosson, Fierville, Normant & Billard 1997; sea bass, Fauvel, Suquet, Dreanno, Zonno & Menu 1998a). Among the criteria used to estimate sperm motility, the percentage of motile cells is significantly correlated with the fertilization capacity of sea bass and turbot spermatozoa.

Because no significant correlation was reported between fertilization rate and the velocity of spermatozoa in these species, the success of fertilization was assumed to depend on the number of motile cells rather than on the quality of sperm movement (Dreanno 1998). However in cod (Gadus morhua L.), the proportion of spermatozoa exhibiting a progressive movement accounted for only a small percentage of the fertilization success (Trippel & Neilson 1992), and spermatozoa showing non progressive vibrating motion could fertilize ova of this species. As a consequence, complementary tests must be used to assess the quality of frozen-thawed spermatozoa.

Assessment of energetic status: the determination of intracellular ATP concentration has proven to be informative in freshwater fish species, because a drastic decrease of nucleotide content has been reported in frozen-thawed spermatozoa compared to fresh sperm (Ogier de Baulny, Le Vern, Kerboeuf, Heydorff & Maisse 1996). Furthermore, a significant positive correlation was reported between the ATP content of turbot spermatzoa and the percentage of motile cells. In sea bass, Adenylate Energy Charge (AEC) is correlated with the percentage of motile cells and the fertilization capacity (Dreanno 1998). The measurement of intracellular ATP content at various times post-activation and the assessment of oxygen consumption allow to assess the degree of cell integrity before and after the cryopreservation (Dreanno et al. 1997).

Seminal fluid composition: in rainbow trout (Oncorhynchus mykiss) seminal plasma pH and osmolality have been selected to determine the semen suitability for cryopreservation (Malejac, Loir & Maisse 1990; Lahnsteiner, Berger, Weismann & Patzner 1996a). A high content of 42kd protein in the seminal fluid was also correlated with a low fertilization capacity of frozen-thawed rainbow trout spermatozoa (Maisse, Pinson & Loir 1988). The oxaloacetic transaminase activity of the seminal fluid of thawed rainbow trout spermatozoa was also considered as a quality test (Malejac et al. 1990). The leakage of these different components into the seminal fluid reflects the alteration of the sperm membrane during the cryopreservation.

Swelling tests: This technique derives from rainbow trout. Sperm samples were incubated in hypo-osmotic solutions and their resistance to the osmotic shock correlated with the cryopreservation ability (Malejac et al. 1990). However, this test is difficult to use in turbot, because no lysis of spermatozoa was observed after dilution in freshwater (Dreanno 1998).- Fertilization capacity: because of the difficulty to collect good quality ova, the fertilization capacity of frozen-thawed spermatozoa has only been tested in a few studies. However, it represents a major step forward in the management of gametes. A standardized experimental insemination procedure is necessary for reproducable evaluation of the fertilization capacity (Suquet, Gaignon, Quemener & Normant 1995a). The problem of replicate variation is illustrated by the sampling technique of ova. When a standard pipette was used, more ovarian fluid than ova was sucked in, resulting in a large methodological error (coefficient of variation: 66.5% in nine successive 800μl samples) in the number of sampled ova. This variation was largely reduced using a positive displacement pipette (5.1%). Weighing ova samples may also be useful. However, correlations between the number of ova and their weight must be calculated for each species and variations could also be due to individual fish conditions or seasonal changes.

The fertilization capacity of frozen-thawed spermatozoa has to be tested using discriminating conditions. In rainbow trout, a significant correlation was reported between fertilization rate and motility estimates using a low number of spermatozoa per ovum (<200,000), whereas a supraoptimal ratio did not allow to detect this correlation (Moccia & Munkittrick 1987). Therefore, knowledge of the optimal sperm to ovum ratio is necessary to improve the cryopreservation technique. Optimal sperm to ovum ratios are only reported for few marine fish species in the literature.

DEVELOPMENT OF THE CRYOPRESERVATION PROTOCOL

To establish a cryopreservation protocol, the extender, the cryoprotectant as well as the cooling and thawing conditions must be investigated. This is difficult as all parameters may interact with each other.

Extender

Pure semen is usually not suitable for freezing. The composition of the most successfull extenders used in marine fish species is reported in Table 2. Except for the ocean pout (Macrozoarces americanus L.), which is an internally fertilizing marine teleost and for which a diluent mimmicking seminal plasma is used (Yao, Crim, Richardson & Emerson 1995), most diluents used in marine fish are saline (concentration 1-10%) or sugar (5-10%) solutions. As motility depends on internally stored ATP which can be resynthesized only at very low rates, the extender must inhibit sperm motility before freezing. However in turbot, addition of cryoprotectants such as DMSO increased osmolality up to 1100 mOsmol/kg and induced the movement of spermatozoa for a period of less than 1 min.

This activation before freezing did not affect sperm motion capacity, probably because turbot spermatozoa are able to resynthetise energy during the period of sperm motion (Dreanno et al. 1997). The Mounib extender (Mounib 1978) is suitable for the cod, sea bass and turbot spermatozoa. The presence of reduced gluthatione prevents deleterious effects of free radicals (Maracine & Segner 1998) and avoids peroxidation of lipids, protecting the sperm membranes (Ogier de Baulny 1997). The Mounib diluent was also successfully used for spermatozoa of freshwater fish (Mounib, 1978; Legendre & Billard 1980). But, Mounib's medium was not appropriate for freezing Atlantic croaker shwater fish species (Billard 1983) and in turbot (Chauvaud, Cosson, uquet & Billard 1995).

(Micropogonias undulatus L.) spermatozoa (Gwo, Strawn, Longnecker & Arnold 1991). Dilution ratios of sperm in extender range from 1: 1 to 1: 20 (volume of semen: volume of diluent). Lower survival of frozen-thawed spermatozoa was recorded for dilution ratios larger than 1: 20 in Atlantic croaker (Gwo et al. 1991) and larger than 1: 50 in seabream (Chambeyron & Zohar 1990). The motility duration of black grouper spermatozoa decreased from 40 to 2 minutes when increasing the semen dilution ratio from 1: 10 to 1: 100 (Gwo 1993). Increasing the dilution rate from 1: 1 up to 1: 9 did not modify the percentage of motile frozen-thawed turbot spermatozoa (Dreanno et al. 1997). It is suggested that seminal plasma proteins protect sperm viability and higher dilution ratios than 1: 10 may reduce this effect. This was observed in fre propylene glycol, glycerol, DMSO and methanol were tested for cryopreservation of spermatozoa of marine fish. DMSO generally gave the best results and its success can be explained by the fast penetration into spermatozoa and by its interaction with the phospholipids of the sperm membrane (Ogier de Baulny et al. 1996). Flow cytometric analysis revealed a high percentage of turbot spermatozoa presenting no cryo-injuries of the plasma membrane and mitochondria in the presence of DMSO (Ogier de Baulny 1997). However, DMSO is toxic at high concentrations: the motility duration of frozen-thawed barramundi (Lates calcarifer, Bloch) spermatozoa was reduced when the DMSO concentration was higher than 5% (Leung 1987) and also in the black grouper, sperm motility was decreased at a concentration of 30% (Gwo 1993).

Methanol has low (barramundi: Leung 1987; turbot: Dreanno et al. 1997) or no (Atlantic croaker: Gwo et al. 1991; black grouper: Gwo 1993) cryoprotective efficiency in frozen spermatozoa of marine fish. Ethylene glycerol and propylene glycerol gave intermediate results in Atlantic croaker (Gwo et al. 1991) and yellowfin seabream (Gwo 1994). In striped trumpeter (Latris lineata Schneider) and in yellowfin bream (Acanthopagrus australis Guenther), the post-thaw motility of spermatozoa was higher with glycerol than with DMSO (Thorogood & Blackshaw 1992; Ritar & Campet 1995). Glycerol gave good protection in yellowfin seabream (Gwo 1994) but it provided no protection in black grouper (Gwo 1993) and low protection in the turbot (Dreanno et al. 1997). Therefore and because of the species specific cryopreservation requirements, the suitability of each potential cryoprotectant must be evaluated when developping a method for new fish species.

In freshwater fish, non-permeating cryoprotectants such as proteins (BSA) or lipoprotein (egg yolk) have been commonly used to prevent damages to the plasma membrane (Scott & Baynes, 1980). Cabrita, Alvarez, Anel, Rana & Herraez (1998) suggested that they increase the membrane resistance to osmotic stress and the motility rate of frozen-thawed rainbow trout spermatozoa. In marine fish, in the Pacific herring, addition of 10% hen's egg yolk did not improve the survival of frozen-thawed spermatozoa (Pillai, Yanagimachi & Cherr 1994). No significant difference was observed in the motility of frozen-thawed turbot spermatozoa when egg yolk (10%), BSA (10%) or a combination of both (5% + 5%) was used (Dreanno et al.

1997). Furthermore, increasing the egg yolk concentration from 10 to 20% or substituting egg yolk with milk, did not change the survival of frozen-thawed spermatozoa in this species (Chereguini, Cal, Dreanno, Ogier de Baulny, Suquet & Maisse 1997). Permeating and non-permeating cryoprotectants can interact: in barramundi sperm, the protective ability of skim milk gave better cryoprotection than egg yolk when glycerol or methanol was used instead of DMSO (Leung 1987).

Because spermatozoa are small, the penetration of cryoprotectants is rapid (Jamieson 1991), and no equilibration period is required. Therefore the toxic effect of DMSO can be minimized. Increasing the equilibration period from 5 to 60 minutes and the DMSO concentration from 10 to 30% lowered the post-thaw motility of yellowfin seabream spermatozoa (Gwo 1994). In seabream, the fertilizing ability of frozen-thawed sperm decreased in DMSO extender when the equilibration period exceeded two minutes (Billard 1978). A similar effect was observed after one hour in grey mullet (Chao, Chen & Liao 1975).

An equilibration time of 10 to 60 minutes had no effect on the fertility of frozen-thawed spermatozoa of hirame (Tabata & Mizuta 1997). In rainbow trout, Ogier de Baulny (1997) observed that DMSO needed 10 minutes to penetrate into spermatozoa althought the fertilization capacity of frozen-thawed spermatozoa was the same with or without equilibration period. This observation suggests that the protective role of DMSO does not depend on its penetration into spermatozoa. With glycerol as cryoprotectant, the motility rate of frozen-thawed spermatozoa of bluefin tuna (Thunnus thynnus L.) was increased at long equilibration periodes (30 minutes) compared to short ones (10 minutes), but the opposite was recorded for DMSO (Doi, Hoshino, Taki & Ogasawara 1982). Since penetration of glycerol is slow, an equilibration time may be necessary.

Cooling Rate

For freezing of fish semen, generally a two step procedure is applied: milt is cooled in nitrogen vapour on a floating tray or in the neck of a container and then the straws are plunged into liquid nitrogen. The cooling rate is determined by the height of the tray or the depth at which canisters are placed. The cooling rates most frequently used in marine fish are shown in Table 4. In Atlantic halibut (Bolla, Holmefjord & Refstie 1987), freezing has also been carried out by pelleting diluted semen onto dry ice. This approximately corresponds to a cooling rate of 35°C.min-1 (Stoss & Donaldson 1982).

Motility of thawed barramundi spermatozoa was not affected by a freezing rate between 1 and 30°C.minute-1 (Lo unpublished result), suggesting high freezing tolerance. On the other hand, a lower freezing tolerance was recorded in seabream as it was optimal only at 10 °C.minute-1 compared to 1, 5, 20, 50 and 100 °C.minute-1 (Billard 1978). In Atlantic cod, a freezing rate of 5 °C minute-1 was successfully applied, whereas 1 °C.minute-1 resulted in low post-thaw motility (Mounib, Hwang & Idler 1968). Freezing conditions for turbot were optimal at a cooling rate of 99 °C.minute-1, but at 46 °C.minute-1 and 148 °C.minute-1 the post-thaw motility rate assessed at 10 seconds post-activation decreased for 8 and 30 % respectively (Dreanno et al. 1997). As a consequence, the optimal cooling rate is highly dependent on the fish species.

Thawing Rate

Rapid thawing is necessary to avoid recrystallisation. Thawing rates used in marine fish are shown in Table 5 and are lower than those reported for freshwater fish (30-80°C: Rana 1995a). In striped trumpeter, there was no difference in the percentage of motile frozen-thawed spermatozoa using thawing temperatures between 10 and 30°C (Ritar & Campet 1995). In turbot thawing temperature between 20 and 40°C did not affect post-thaw sperm motility (Dreanno et al. 1997). But, an increase in the thawing temperature from 1 to 30°C reduced the post-thaw motility of ocean pout spermatozoa (Yao et al. 1995). When Atlantic croaker spermatozoa were thawed at 0°C, the fertilization rate was significantly lower than at 25 and 50°C (Gwo et al. 1991). In the Atlantic halibut, the optimal thawing temperature was 10°C for straws and 40°C for pellets (Bolla et al. 1987).

Thawed spermatozoa must be rapidly used: 60 minutes after thawing, the percentage of motile turbot spermatozoa stored on crushed ice decreased for 35% (Dreanno et al. 1997). This indicates that cryopreservation induced damages in spermatozoa. Diluting frozen-thawed spermatozoa of this species in a medium mimicking the seminal fluid improved their short term storage capacity. Also, in halibut, short term storage ability of thawed sperm was lower than for fresh sperm (Billard, Cosson & Crim 1993).

The Quality of Frozen-thawed Spermatozoa

Generally, the motility rate of frozen-thawed spermatozoa is very high in marine fish species when compared to freshwater species: 21% in rainbow trout (Lahnsteiner, Berger, Weismann & Patzner 1996b), 25% in carp (Babiak, Glogowsky, Brzuska, Szumiec & Adamek 1995), and from 40 to 85% in tilapias (Chao, Chao, Liu & Liao 1987). In marine species, the high motility rates of frozen-thawed sperm result also in high fertilization rates as recorded in sea bass (fertilization expressed as a percentage of fresh semen control: 65%; Fauvel et al. 1998a) and turbot (83%; Dreanno et al. 1997), using limiting quantities of sperm. The decrease in the fertilization capacity of frozen-thawed semen probably reflects the changes in motility rates observed after freezing and thawing. Other parameters describing the movement of thawed spermatozoa are rarely assessed. In sea bass, the flagellar beat frequency of swimming spermatozoa was not affected by the cryopreservation process, but straight line and curvilinear swimming velocities were significantly decreased (Fauvel et al. 1998a). The straight line velocity of frozen-thawed turbot spermatozoa did not significantly differ from that of fresh sperm (Dreanno et al. 1997).

Cryo-injuries have been reported for thawed spermatozoa of many freshwater fish species (Rana 1995a). In marine fish, the fine structure of the head region of 90% of the thawed black grouper spermatozoa was similar to untreated ones (Gwo 1993). Also in the puffer, 80% of the frozen-thawed spermatozoa had similar ultrastructure as untreated ones (Gwo, Kurokura & Hirano 1993).

Shrinkage of the plasma membrane of the midpiece was reported for frozen-thawed spermatozoa of the ocean pout (Yao et al. 1995). In frozen-thawed Atlantic croaker spermatozoa, the cristae of mitochondria were disrupted, plasma membrane was swollen or disrupted and the axoneme coiled (Gwo & Arnold 1992). Flow cytometric analysis of frozen-thawed turbot spermatozoa revealed a high percentage (up to 93%) of intact plasma membranes and mitochondria (close to 80%; Ogier de Baulny 1997) and the mitochondrial respiratory activity was not altered (Dreanno et al. 1997).

The development of fish produced with frozen-thawed sperm was not frequently investigated. Fertilizing ova with cryopreserved sperm did not affect the hatching percentages in cod. Furthermore, the morphology of larvae produced with frozen-thawed spermatozoa was similar to those produced with fresh sperm (Mounib 1978).

Also, the rate of malformed hirame larvae was not significantly different when using fresh or frozen-thawed spermatozoa (Tabata & Mizuta 1997). In the yellowtail flounder, the hatching rate and the percentage of malformed larvae were not affected by the cryopreservation process (Richardson, Crim, Yao & Short 1995) and also 29 days old barramundi larvae, produced with frozen-thawed spermatozoa showed no malformalities (Palmer, Blackshaw & Garrett 1993).

The hatching rate, the larval survival rate and the larval weight of 10 days old turbot were not significantly different when ova were inseminated with fresh compared to frozen-thawed sperm (Suquet, Dreanno, Petton, Normant, Omnes & Billard 1998b). But, the hatching rates of eggs inseminated with frozen-thawed sea bass sperm (69%) were significantly lower than those obtained with fresh sperm (81%) (Fauvel, Zonno, Suquet, Storelli & Dreanno 1998b).

Compared to fresh water fish, the high quality of cryopreserved spermatozoa of marine fish species has already been emphasized by Scott & Baynes (1980). Ogier de Baulny (1997) found a significant correlation between the percentage of motile frozen-thawed turbot spermatozoa and the membrane integrity as determined by flow cytometric analysis.

Furthermore, this author recorded that up to 93% of frozen-thawed spermatozoa had intact plasma membranes in the turbot, 45% in the trout, 80% in tilapia and 90% in catfish. Drokin (1993) proposed that the cryoresistance of marine fish spermatozoa could be due to the lipid composition of sperm membranes, mainly to the molar ratio of cholesterol to phospholipids which is 2-3 times higher than in freshwater fish. Contrary in rainbow trout spermatozoa, a lower cholesterol content of the plasma membrane was correlated with a higher fertilizing capacity of frozen-thawed sperm (Labbé & Maisse 1996), and the ratio of cholesterol to phospholipids was not higher in turbot than in trout (Ogier de Baulny et al. 1996). According to Drokin (1993), phosphatidylcholine may increase the cryoresistance of the sperm membrane and in sperm of marine fish this component has higher levels than in freshwater fish. Its protective role against osmotic and cold stress has been also reported by Simpson, Swan & White (1986).

The process of cryopreservation reduces the decrease of intracellular ATP content of turbot spermatozoa for 20-40% (Ogier de Baulny 1997; Dreanno et al. 1997). This decline is lower than in trout (50-90%) and similar to catfish (25%; Ogier de Baulny 1997). This low rate of ATP consumption could explain the higher motility percentages observed in frozen-thawed spermatozoa of turbot (70%) and catfish (60%) compared to trout (21%). Furthermore, the significant decrease of cryoresistance of sea bass spermatozoa recorded at the end of the milting season could be explained by the decrease in endogeneous ATP (Dreanno et al. in press). Contrary, ATP content decreased only slightly in tilapia, Oreochromis niloticus (L.) spermatozoa during cryopreservation (0-25%).

However, only 22% of the frozen-thawed spermatozoa of this species could be activated (Ogier de Baulny 1997). Before movement, ATP content of frozen-thawed trout spermatozoa is about 1.85×10^{-2} mmole/10^9 spermatozoa.

When calculating the internal cell volume as 0.16 μl/10^7 spermatozoa (Christen, Gatti & Billard 1987), this leads to a 1.2 mmole/l concentration. According to Saudrais, Fierville, Cibert, Loir, Le Rumeur & Cosson (1998), the demembranated trout spermatozoa move also at much lower concentration of ATP (0.2 mmole/l).

Therefore, the lower losses of intracellular ATP levels can not explain the higher survival of frozen-thawed turbot spermatozoa compared to trout spermatozoa. Complementary studies on the cryoprotectant toxicity and on thermal and osmotic stress should be carried out in marine fish species to understand their high capacity to tolerate the cryopreservation.

CONCLUSIONS

Simple freezing protocols are available for the cryopreservation of marine fish semen. The composition of the extender is generally less complex than the seminal fluid and consists of DMSO containing saline or sugar solutions. Simple methodology as straws on a tray in the vapour of liquid nitrogen and thawing in a bath are used for freezing and thawing.

To date, the semen of about 30 different marine fish species has been cryopreserved and compared to freshwater species, a high survival of frozen-thawed spermatozoa is often recorded. In the literature, the species specific differences in cryoresistance are related to species specific differences in ATP consumption and to various grades of damages of frozen-thawed spermatozoa especially of their mitochondria and plasma membranes.

However, studies on the cryopreservation of spermatozoa of marine fish species were mostly concentrated on the improvement of freezing technique protocols, but lacked a complete description of morphological and metabolic changes. Also, the quality of sperm sample before freezing should be investigated in more details as the problems of semen aging and of urine contamination can alter the biological features of spermatozoa and their suitability for freezing.

The possible improvement of sperm fitness for cryopreservation by modifying rearing parameters during spermatogenesis (e.g. water temperature and food composition) has not yet been studied in marine fish species. The effect of the process of domestication on the quality of sperm and its ability to be frozen has also not been described. Therefore, the cryopreservation techniques for semen of marine fish are applicable for production purposes in aquaculture and for establishement of sperm banks. Coupled with insemination and short term storage techniques, cryopreservation will lead to an improvement of gamete management in marine fish species.

HORMONAL MANIPULATION OF FISHES AS AN AQUACULTURAL TECHNIQUE

Aquaculture is the controlled reproduction and growing of aquatic organisms for food or ornament. Current trends in human population growth, habitat and natural stocks destruction versus advances in world agricultural and fisheries production clearly show the basic need for developing and implementing aquacultural technology.

Widespread use of hypophysation (use of the pituitary gland, a part of the brain and endocrine system of humans and fishes) has permitted breeding of a number of species otherwise not spawned in confinement (e.g. kuhli loaches, Acanthopthalmus kuhlii, dolphinfish, Coryphaena hippurus).

Many of the important cultured pet and food fishes fall into this category. Previously fingerling-stock had to be collected from the wild. This supply was seasonal and unpredictable. By manipulating hormonal and environmental factors seed can be produced all year long; reducing cost, disease, and incidence of trash fish (undesireable contaminants) and predators. (Hickling 1968, 1971; Jhingran 1969).

Hypophysation involves the injection of gonadotropins (hormones=chemical messengers produced by the pituitary, that stimulate the reproductive organs) from one animal into another. Among vertebrates (fishes, amphibians, reptiles, birds and mammals) these are not very species-specific. Sometimes it has been found better to mix mammalian pituitary extracts and steroids in varying concentrations with fish gonadotropin. There is increased success in the use of prepared mammalian hormones in breeding procedures, preservation of fish sperm, maintenance of brood stock; yielding reliable sources of fish seed and acceleration of breeding research. Problems relating to standardization of dosage, determination of maturity of recipient fish and nutritional requirements particularly need to be further explored. Current knowledge on the use of pituitary hormones in fish culture and technology regarding collecting, processing & storing of glands & methods of application have been covered comprehensively in the literature.

S. Aschheim & B. Zondek are credited with the first experiments regarding reproductive behavior (in female mice) and replacement therapy using pituitary implants. The first use in fishes was in 1930 by B.A. Houssay of Argentina. He induced premature birth in a viviparous (livebearing, like guppies) fish by injecting pituitary glands recently removed from other fish. Other South Americans, like von Ihering, a Brazilian, used this technique for aquaculture extensively beginning in 1934.

Soviets attempted to induce captive sturgeons to spawn in 1932 by means of mammalian hormones.

They were unsuccessful until 1937, when N.L. Gerbil'skii produced eggs and sperm in a number of the sturgeon, Acipenser stellatus that had been intracranially injected with 1-2 fresh pituitaries of conspecifics (the same species). As of 1957 the Soviets have obtained all their sturgeon eggs for culture from pituitary treated fishes, very important to their caviar industry. As of 1963 the Chinese have been able to supply all the seed used in carp culture (big-head, silver, common & other food species) by means of hypophysation.

METHODS OF INDUCED BREEDING

Fish are induced to spawn by altering environmental parameters, particularly temperature and photoperiod, and hormonal constitution. These variables are generally constant for the species but vary widely among cultured food fishes.

Environmental Factors & Diet

Environment and diet are the principal sources of phenomena determining hormonal secretion. Several factors, photoperiod, temperature, metabolites, pheromones, light strength and temperature shock control release of gonadotrophins, which in turn induce gamete (eggs & sperm) production and concomitant reproductive behavior. Reproduction may occur out of season by manipulation of temperature and photoperiod (how long the light is on daily) alone.

An example of improved culture just through manipulation of environmental factors may illustrate the point. The Ayu, Plecoglossus altivelus, a smelt-like fish cultured in Japan during the summer months, normally spawns in October-November., after which most of the fish die. Due to lower water temperatures that time of year, it's difficult to maintain live food organisms in adequate densities in rearing ponds to feed fry. Adjustment of photoperiod resulted in successful acceleration of sexual maturation (August), enabling rearing of fry when natural food was abundant. Spawning was also delayed to prolong the life of adults. The light period was extended to 18 hours per day with artificial light (August through October). Reproduction was retarded & adults were marketed out of season in February.

Temperature

The relative importance of temperature varies per species, investigator and experiment, but all agree a certain range and lowering or elevating optimizes results. Smiglieski (1975) stated that temperature was the controlling factor in his work with flounders. Yamamoto et. al (1966) observed spermatogenesis is goldfish, Carassius auratus below 14 degrees C & accelerated spermiation above 20 degrees C. Often, if much temperature fluctuation occurs, reproductive behavior will cease.

Haydock (1971) has observed temperature threshold below which gulf croaker will not hydrate or ovulate. For each species or sometimes race (a sub-specific classification) there are limits, both high and low, which eliminate reproduction physiologically and behaviorally.

Light

As with temperature, there are optimal amounts of strength, quality and duration of light in relation to reproductive behavior. On work with bluegill sunfishes it has been found that a longer photoperiod 16light/8dark versus 8L/16D induced a better gonosomatic index (G.S.I.= ovary weight divided by total body weight times 100). Males were more aggressive, dug more nests and fertilized more eggs. Hoar (1965b) sums up the effect of longer light periods by stating that they stimulate secretory activity of anterior pituitary, inducing pre-sexual behavior. The pituitary LH, Lutenizing Hormone (one of the gonadotropic hormones produced in the brain) activates interstitial tissue of the gonads which produce gonadal steroids that in turn dominate sexual phases, taking complete control during prenatal phases. Several investigators have shown that elevated temperature and prolonged photoperiod increased gonad maturation in the green sunfish, Lepomis cyanellus.

Other

Photoperiod and temperature have received most attention & are generally considered to be of greatest importance in inducement of sexual maturity and spawning. Other factors such as the effects of other environmental stimuli; meteorological (rain, floods, etc.) & water conditions (pH, Ammonia, C02, turbidity) on controlled breeding of food and ornamental fishes has been investigated. In several species, the presence of con-(the same species) and/or hetero-(other) specifics is important. In bluegills and some cichlids it has been found that when males are present, the most aggressive female has greaater ovarian development. Another application of environmental control is employed in mullet culture in Israel. By keeping the adults in freshwater ponds and not allowing access to the sea where they migrate to spawn, fish-culturists are able to extend the spawning season.

Diet

Researchers have demonstrated gonadal regression induced by restricted diets accompanied by an inversion (opposite) of the lasophil/acidophil (cell types distinquished by their staining characteristics) ratio of the mesoadenohypophyis (an anterior part of the pituitary gland in ours and fishes brains) and a reduction in gonadotrophin content. These conditions were reversed with adequate food. Some authors have stressed the interaction between photoperiod and diet and pointed out that optimum benefit from photoperiod adjustment can be over-rided by poor diets. Hmmm.

HORMONAL MANIPULATION

Criteria: Biological and Chemical Assay

Several methods are applied towards determining sex and sexual maturity of spawners. Externally there may be sexual dimorphism (structural differences between the sexes) in the species and/or differences that occur during sexual ripeness. Such is the case with the channel catfish, Ictalurus punctatus. Females assume a dropsical appearance with raised and reddened genitalia. The males develop swollen head musculature and well defined genital papilla.

A practical and precise way to measure oocyte (egg cell) development is through inserting a sterile polyethylene or glass catheter in the oviduct and removing egg samples. The eggs are cleared in successive concentrations of ethyl (grain, drinking) alcohol (30, 50, 70, 90, 100%) then xylene (an organic solvent) for 5 minutes each. Eggs are removed by catheter before each hormone injection to determine gonadal development.

Physiological parameters are also valuable as sources for assays. Elevated plasma (the fluid portion of blood, the same as for humans) calcium levels have been reported during gonad development in many species. Peterson & Shehadeh (1971) reported a four time increase in calcium and two time plasma copper in striped mullet, Mugil cephalus, under methyl testosterone treatment. For examination, gonads are removed from the body cavity, weighed to 0.01 gram, labeled, wrapped in cheese cloth, fixed in 70% ethanol (ethyl alcohol) for 30 minutes and 100% for one hour. Xylene is used for a clearing agent. Infiltration and final embedding is done in fresh paraffin (wax). Typically microscope slides are produced by slicing these specimens on a tool called a microtome at 8 micron thickness. These tissue slides are stained with a compound called Delafields iron hematoxylin and counter stained (for contrast) with eosin. The specimen is permanently covered with a thin glass coverslip and permount.

When you examine these prepared sections under a microscope you can judge the sexual readiness oth your stock. In sexually immature males the lumen (opening) of the seminiferous tubules is unformed. Later primary and secondary spermatocytes develop in the lumen. Still later spermatids and spermatozoa (these are all developmental phases of male reproductive cells) are seen free in the lumen. After spawning the tubules are completely occluded (blocked) and testes become highly vascularized (serviced with blood vessels). In immature females only the oogonia & primary (previtellogenic) oocytes are present. At maaturity secondary oocytes distributed by oolemma and centripedal yolk precursor form in the cytoplasm (the fluid portion of cells).

The pituitary may be examined in sample individuals. They are dissected and promptly fixed in Bouin's or Zenker's for 24 hours. They are subsequently passed through alcohols and stored in 70% ethanol, further processed and sectioned.

In the channel catfish the adult pituitary double from one to two millimeter maximum length at sexual ripeness. The mesoadonhypophysis contains gonadotropic and somatotropic cells; the former enlarge in ripe specimens. In this way the mesoadenohypophysis becomes large during sexual maturity.

Problems of Standard Dosage

The endocrine (=cry internally) system acts like a chemical link between an organism and it's environment. Fluctuations in the external environment and diet mediated through the central nervous system (CNS) affect pituitary secretion. As one of several target tissues the gonads are effected. The CNS acts quickly, temporarily and specifically. The endocrine works in concert and antagonistically with the CNS but slower, longer-lasting and more generally.

It is well established that in all vertebrates the pituitary gonadotropins and gonadal steroids are the hormones most directly involved in the regulation of reproductive behavior. Pituitary hormones may act by stimulating secretion of gonadotropic hormones, which in turn regulate behavior. Also, individually these hormones may influence behavior indirectly by their role in the development of secondary sexual characteristics.

In turn the CNS-pituitary-gonad axis is influenced by internal and external factors acting through the sense organs and brain. Studies on birds have emphasized that coordination and instigation of components of reproductive behavior are dependent on a continuous interplay of external and internal factors; hormones effect behavior and behavior effects the endocrine system.

There is some evidence that hormones other than those of the anterior pituitary and gonads play a major role in reproduction. Neurophysiological and thyroid hormones most outstandingly, but other hormones affect reproduction indirectly by their effects on general metabolism and growth. All of the endocrine glands are either directly of indirectly involved, since reproduction requires profound metabolic adjustments, particularly in females.

Injections: Two methods of injection are in wide practice. Intramuscular, in the flank just below the dorsal fin and behind the gill cover. This method is safer but slower working. Interperitoneal injections are made into the body cavity. They are faster acting but involve a greater chance of injury or death.

Dosage, latency period and frequency of injection have not been well determined for any one species; investigators adjust variables according to specific circumstances. This has yielded divergent results, duplication of effort and failure. Problems of determining exact dosages lie in two factors. 1) the largest fish does'nt necessarily need the largest dose; this is more dependent on the hormonal development of the individual. 2) Exact dose of any preparation depends on determination of it's gonadotropic content. Towards the ends of standardization an international standard animal for bioassay has yet to be agreed upon. The Soviets use the Vy'un unit (U.V.) based on the loach, Misgurnis fossilis, & the frog unit (Curimata sp.).

The Brazilians use the fish unit (Fontenele 1955). A Wagur unit using Clarias batrachus (the walking catfish!) was advocated by Das & Khan (1962). In the U.S.A. changes in the gonadosomatic index (GSI) of hypophysectomized (with pituitary gland removed) cyprinodontid (killifish family) Fundulus heteroclitus have been often used. More recently, a bioassay assays using seminal fluid viscosity have been used. Goldfish spermiation was used by Yamazaki & Donaldson (1968) in response to purified salmon gonadotropin. Otsuka (1956) monitored uterine and ovarian changes in immature mice. The Galli-mainini reaction and spermiation or hydration in goldfish has been used most frequently. For convenience the wet weight of pituitary or dry weight (acetone dried) of pituitary powder/ weight of fish is unfortunately still used versus chemical analysis of hormones to determine which are present, active and their synergistic-antagonistic effect on each other.

So it can be seen that due to the variety of test animals as standards, & lack of test for definitive presence and concentration of hormones, comparison of results is difficult, and therefore the data are mostly of local benefit.

More precise bio-physico-chemico-assay methods such as incorporation of radio-phosphorus by gonads is in it's infant stages. (This is a specific labeling and measuring technique)

Maturity of Recipient Fish

Variation in a species in terms of gonad development is intensified in confinement. Variability of results is due to individual variation in sex or season, especially considering crude bioassay techniques. Effectiveness of hypophysation (injection with pituitary hormones) is dependent on the stage of reproductive development of recipients. Aquaculturists recommend that when applying other workers' data; that matching the dosage to the physiological state of the recipient can best be done by administering doses in graded series from above or below those recommended by other investigators. Almost always slightly greater success is achieved with the most intense regime (injecting less hormone more frequently).

If the species has not been spawned before they suggest continuous injections, every 24 hours, with a wide range of doses until spawning does occur. (This is science?)

As has been mentioned, care must be exercised in assaying sexual readiness in spawners. Parameters generally adopted have frequently proven unreliable. An example of this is females with enlarged abdomens, reddish coloration & protrusion of the cloacal region may be due to engorgement of the intestine, or disease, even during the spawning season.

Variability in Pituitary Gonadotropin Related to Phylogeny, Sex & Season

In order to determine dosage, effect and success of treatment it is vital that fluctuations in gonadotropic activity related to phylogenetic specificity (basically how closely species are related to each other evolutionarily), sex and season be known for each species. Much controversy exists as to specificity in gonadotropin as it relates to vertebrate phylogeny.

Work with homoplastic (derived from the same species) pituitaries has produced best results. Little correlation exists between phylogenetic affinity and effect of pituitary gonadotropins. Where distantly related donor-recipients are used the general practice has been to increase dosage.

The early literature on sex related differences in the gonadotropic activity of fish pituitaries has been outlined by Pickford & Atz (1957). Much of it and later reports are contradictory. It is more or less agreed that sex-related differences in pituitary extracts are usually minimized by the usual preparation of mixed extracts of both sexes in excess of threshold requirements. Seasonal fluctuations in gonadotrophic activity are not uncommon in almost all fish species examined. Pituitaries collected for hypophysation are almost always taken from mature fishes during spawning season to be used immediately or dried and prepared for later use. (Kind of hard on your breeders, eh?)

Use of Mammalian Hormones

Advantages of using mammalian hormones include:

(1) Reliable source, available all the time.

(2) Easily stored in lab and retain biological activity for a number of years.

(3) More uniform than crude pituitary preparations, permitting controlled experiments and repeatable results.

(4) Cost is about the same.

(5) Saves valuable gravid fish, especially where homoplastic pituitaries are used.

(6) Does away with the tedious task of collecting, preserving, processing & distributing pituitary material.

There is not much in the literature on the use of such hormones for practical breeding of ornamental fishes; quite a bit on food species. Several Indian workers have worked extensively with Heteropneustes and mammalian hypophysial hormones, placental gonadotrophins, gonadal hormones and corticosteroids finding that human chorionic gonadotropin (HCG) and some of the corticoids, especially deoxycorticosteroids could be used profitably to spawn food fishes as a substitute for pituitary extracts. Chorionic gonadotrophin was also found to be an effective ovulating agent in Clarias batrachus. Snead & Clemens (1959) established methods for spawning channel catfish, goldfish, and white crappie, Pomoxis annularis with injections of chorionic gonadotrophins. At the price of $5.65 per 10,000 international units (IU) of APL (Anterior Pituitary Like) they established a cost of 45 cents per pound of fish spawned, comparable to established costs of fresh pituitary preparations.

Other reviews of successful use of mammalian hormones exist detailing one's effectiveness over another, acting singly and together.

Purified fish gonadotropins are available to the fish breeder, especially fractionated salmon pituitary. Much work has been done with these and food fishes; not much reported for pet fish.

Pituitary Extracts, Use & Preservation

In a fish, the pituitary is concerned with regulation of growth, pigmentation, adrenal cortex, thyroid gland and others, as well as reproduction. Cyclical changes in the pituitary are striking, showing increase and decrease in cyanophil cells involved with production of gonad stimulating hormones.

The synergic action of growth hormone, thyrotropic hormone (stimulating the thyroid gland) & sex steroids is indicated by the better results obtained with crude rather than purified gonadotropic hormones.

The pituitary regulates both gametogenesis and steroidogenesis in fishes. Two seperate gonadotropins Follicle-stimulating hormone (FSH) and Lutenizing hormone (LH) are physiologically distinct in the tetrapods (= "four legs", the rest of the vertebrates beyond the fishes, i.e. the amphibians, reptiles, birds & mammals). There is evidence of a single protein in fishes. When fish pituitary extracts are tested in tetrapods, both FSH and LH effects are elicited.

With rare exceptions induced spawning by hypophysation has been restricted to females. This is often due to the availability of male seed.

Collection of pituitaries involves chopping the head off the fish above the eyes, lifting the brain and removing the gland from the base of the cranium. (Ouch! How do you put it back together?) Preserving the gland is done by freezing immediately with dry ice or dehydrating in acetone. This is done by changing the acetone several times every 12 hours. This is then stored in a desiccator (dryer). It lasts up to two years but the effect is not as good as fresh.

When preparing, care must be taken to not introduce foreign proteins. Combined with pituitary this sometimes causes very strong rejection. Therefore purification is necessary. Whole fish pituitaries may be ground up or injected whole. If ground, they are suspended in physiological saline or distilled water.

Gonadal Steroids

In all fishes investigated development of secondary sexual characteristics (are they growing facial hair?) has been shown to depend on gonadal steroids. Several androgens (male sexual hormones like testosterone) and estrogens (the female equivalent) have been identified from the tissues and blood of fishes. Evidence shows that some fish gonadal steroids have the same biochemical pathways as higher vertebrates. Among the steroids found are testosterone, androstenedione, 17 beta estradiol, androsterone, 11 keto testosterone, conjugated testosterone and progesterone. As with higher vertebrates bot males and females have compounds in common but in varying concentrations.

Hoar (1962b) presents evidence that reproductive behavior becomes dependent (in a feed-back loop with the pituitary) on gonadal hormones and less sensitive to pituitary gonadotropic hormones. Acquistion of nuptial coloration & onset of breeding behavior has been correlated with increasing steroidogenesis.

DISCUSSION

Even in a paper of this scope it's possible to draw general conclusions about the role of hormones in concert with environmental factors in control of spawning behavior of fishes. In male fishes of several species, secondary sex-morphological changes are dependent on gonadal hormones. In several species, components of reproductive behavior are under gonadal control, yet in other species much of all reproductive behavior persists after castration. (Is this evidence of learned behavior, conditioned response?)

Ovariectomy abolished sexual response in several species examined. In the guppy sexual behavior reappears some time after ovariectomy. There has been no unequivocal demonstation that administration of estrogen or progesterone will induce reproductive behavior in fishes.

Some of the reasons allowing for the differential success in use of hormones to control reproduction in fishes are the phylogenetic differences in protein and peptide hormones of the pituitary. On the other hand, the similarity of steroids suggests there is little problem in phylogenetic specificity in applying steroid therapy.

I'd like to see controlled studies on the use of steroids on color and growth augmentation in sport mutations of oscars, Astronotus, guppies, and discus, Symphysodon among other hormone-treated species from the far east, in terms of related reproductive viability, longevity and vitality. How about you?

Pet Fish Conclusions: I hope the following became clearer through this review:

(1) The effects of temperature, light (duration, quality & strength), metabolites ("fish wastes"), are very important, and many other factors to a lesser extent, in getting stock up to and increasing success in reproduction.

(2) There is an extensive body of knowledge in scientific literature on the behavior and physiology of fishes, in particular food & game species.

(3) That you probably "knew" much of this material intuitively from your association with aquarium species, and indeed could write a "scientific" paper of substantial importance concerning your experience, if not a thesis.

TEMPERATURE-DEPENDENT SEX DETERMINATION IN HD-RR MEDAKA ORYZIAS LATIPES

The developmental time and thermal threshold for temperature-dependent sex determination (TSD), gender differences in temperature sensitivity, the fertility of thermally sex reversed fish, and the effect of temperature on the expression of two major sex determination/differentiation genes (DMY/DMRT1bY and DMRT1) were examined in the Hd-rR strain of medaka, Oryzias latipes. Fertilized eggs were exposed from either shortly after fertilization (8-16 cells; embryonic stages 5-6) or from middle embryogenesis (heart development stage; stage 36) until hatching to temperatures ranging from 17°C to 34°C. Secondary sexual characteristics, gonadal histology, progeny testing, sex-linked body coloration and gene expression were used to determine phenotypic and genotypic sex. Sex determination was unaffected by low or high temperatures in genotypic (XY) males. In contrast, genotypic (XX) females treated from stages 5-6 showed increasing rates of sex reversal into phenotypic males at temperatures above 27°C up to 100% at 34°C. Thermal manipulation of sex was ineffective after stage 36, indicating that gonadal fate in medaka is determined considerably earlier than histological differentiation (stage 39). High temperature induced DMRT1 expression in genotypic females, which was observed already from stage 36. Sex-reversed males had histologically normal testes, were capable of sexual courtship and, with the exception of fish from 34°C, sired viable progeny when mating with fertile females. These results clarify the pattern of TSD in medaka and provide important clues to understand the mechanism of sex determination in this species. They also suggest that a brief exposure to high temperature early in life could impair the fertility of medaka as adults.

A growing body of experimental evidence suggests that the process of sex determination (or differentiation) in fish is quite sensitive to the effects of environmental temperature [Baroiller et al., 1999; Devlin and Nagahama, 2002; Strüssmann and Nakamura, 2002]. In temperature-dependent (or thermolabile) sex determination (TSD), the thermal conditions during a critical phase in early life can induce the formation of functional ovaries or testes regardless of putative genetic orientation. This phenomenon occurs over a broad range of environmentally relevant temperatures and appears to be not restricted to species from any phylogenetic group, lifestyle or habitat, although it is common and very marked among Atheriniformes [Strüssmann and Patiño, 1995; Strüssmann and Nakamura, 2002]. In this taxon, TSD is indeed the main determinant of the sex ratios in natural populations [Conover and Kynard, 1981; Middaugh and Hemmer, 1987; Strüssmann et al., 2003] and supposedly confers an adaptive advantage to some species (e.g. the Atlantic silverside, Menidia menidia) [Conover, 1984].

However, reports of TSD are available also on dozens of unrelated species, including many which have previously been assumed to have strictly genotypic sex determination (GSD) such as sockeye salmon, Oncorhynchus nerka [Craig et al., 1996], tilapia, Oreochromis niloticus [Baroiller et al., 1995], zebrafish, Danio rerio [Uchida et al., 2004], channel catfish, Ictalurus punctatus [Patiño et al., 1996], and even viviparous species such as Poeciliopsis lucida [Schultz, 1993]. In this context, there appears to be room for the argument that some degree of TSD could very well be the rule rather than the exception among fishes, and that modulation of sex differentiation by temperature probably owes as much to the poor state of development of the mechanisms of sex determination in fish as to poikilothermy itself [Strüssmann and Patiño, 1999]. Moreover, in view of the projected trend of rising average global temperatures and ensuing thermal anomalies, the possibility of effects of temperature on the process of sex differentiation becomes a grave concern to the reproductive health of fish populations.

In medaka Oryzias latipes, the first species of fish in which a main sex determining gene was documented (DMY or DMRT1bY as identified by Matsuda et al. [2002] and Nanda et al. [2002], respectively), sex is generally considered to be strictly determined by the animal's genotype. In this case it is an XX/XY system, as demonstrated by detailed progeny analysis [Aida, 1921; Yamamoto, 1969], cytogenetic analysis [Matsuda et al., 1998], and linkage mapping [Matsuda et al., 2002; Nanda et al., 2002]. Although this sex determining system appears to be fairly conserved within the genus Oryzias, only a few species appear to possess the DMY/DMRT1bY gene [reviewed by Schartl, 2004]. Moreover, naturally occurring XX males have been found in several wild [Shinomiya et al., 2004] and laboratory [Nanda et al., 2003] populations of medaka, suggesting that there might be alternate systems of sex determination in this species. Of interest is the observation of Aida [1936] who reported increased proportions of males during hot summers, but little has been done to explain this phenomenon.

More recently, studies conducted in our laboratory provided preliminary evidence of TSD in medaka in the form of synergism and antagonism between exposure to low and high temperatures, respectively, and a feminizing hormone (estradiol) during the early stages of sex differentiation [Minamitani and Strüssmann, 2003].

We have since conducted a thorough examination of the effects of temperature on the sex determination of this species. While this paper was in preparation, a report of a similar study on the effects of temperature on sex determination of the Hd-rR and HNI strains of medaka appeared [Sato et al., 2005].

That study provided the first clear evidence of TSD in medaka, showing that embryos of genotypic (XX) females of both strains developed as functional males when exposed to a temperature of 32°C between fertilization and hatching.

However, no information was provided on the timing of TSD in relation to sex differentiation and on the effects of other temperatures in this process. Likewise, the molecular mechanism of TSD in this species has not been clarified.

In this study, we present the results of experiments spanning the full range of viable temperatures of embryos (15°C to 34°C) on the sex differentiation of both genotypic males and females of the Hd-rR strain of medaka. We also examined the timing of sex determination by thermal treatment of medaka embryos during two different developmental stages and examined the fertility of sex reversed individuals.

Finally, we examined the pattern of expression of the putative sex-determining gene DMY/DMRT1bY and of the autosomal gene DMRT1, which in medaka is associated with spermatogonial differentiation [Kobayashi et al., 2004], in order to obtain clues on the molecular mechanism of high temperature-induced masculinization in medaka.

MATERIALS AND METHODS

Source of Animals

Medaka from the highly inbred Hd-rR strain were a kind gift from the Animal Supply Facilities, National Institute of Radiological Sciences, Chiba, Japan. To obtain fertilized eggs, we set up nine 20-liter plastic tanks containing 4 adult females and 2 males each. This brood stock was reared in brackish water (0.1-0.4% NaCl) at a temperature of 25°C and a photoperiod of 14L-10D. Fish were fed Tetramin (Tetra®) flakes and supplemented with live Artemia nauplii ad libitum.

Embryo Collection and Incubation

Fertilized eggs were collected daily within 2 h of spawning, and stocked in Petri dishes with brackish (0.1-0.4% NaCl) water and methylene blue (50 ppm) as a prophylactic at a stocking density of up to 50 eggs per dish. To assess the effects of temperature and the thermolabile period, groups of eggs were transferred within 1 h of collection (3 h from spawning, developmental stages 5-6) [Iwamatsu, 1994] to incubators (IUCHI Mini-Incubator UI-50) pre-set at 17, 21, 25, 27, 30, 32 and 34 ± 0.3°C or after a preliminary incubation period of 6 days at 25°C (until the heart development stage, or stage 36) to incubators set at 15, 18, 22, 25, 26, 30, 31, and 34 ± 0.3°C. These experimental temperatures were chosen based on empirical results on the thermal tolerance and hatchability of Hd-rR strain medaka eggs and those of Yamamoto [1975]. After 24 h, unviable eggs were discarded, the chorionic filaments were trimmed off, and the rearing water was exchanged completely for new water previously adjusted to the same temperature. Thereafter, unviable eggs were removed and the water was exchanged every 48 h until hatching.

Sexing of Fish

Hatched larvae were immediately separated into groups according to their incubation temperature and reared under the same standard conditions as for brood stock until reaching sexual maturity at about 2-3 months. No attempt was made to discriminate between larvae from different spawnings or parents within the same temperature incubation group because of the high degree of inbreeding of the brood stock. After attaining sexual maturity, each fish was analyzed to determine its phenotypic and genotypic sex. Phenotypic sex was determined by observation of secondary sexual characteristics in fin morphology and anal papilla under a stereomicroscope following identification criteria of Yamamoto [1975] and Iwamatsu [1999].

Phenotypic sex was also examined by light histological analysis of the gonads in selected individuals to confirm the findings of the observation of secondary sex characteristics. For this purpose, fish were fixed in Bouin's solution for 24 h and preserved in 70% ethanol. The trunk region containing the gonads was then processed for histological analysis by routine histological methods. Sagittal histological sections of 6 μm of thickness were stained with hematoxylin-eosin and examined under a microscope. The genetic sex of each individual was identified by two methods. First, we observed each fish under the stereomicroscope concomitant with the determination of secondary sexual characteristics and scored the genetic sex based on body pigmentation. This is possible in the Hd-rR strain of medaka because inheritance of sex and body pigmentation are linked: genotypic (XX) females are white and genotypic (XY) males are orange-red [Yamamoto, 1975].

Second, animals were subjected to DNA analysis for the presence of the DMY/DMRT1bY gene. For this purpose, we collected a piece of the caudal fin of all individuals sacrificed for histology and extracted their total DNA using the DNeasy® Tissue Kit (QIAGEN). A PCR analysis was then performed using 4 sets of primers (table 1) to cover the DMY gene in its entirety.-Actin was amplified as a positive control in the case of primers EX3.1/3.2, EX6.1/6.2, and DMT1K/DMT1l, since genetic females do not show any band with them. The PCR products were analyzed on a 0.9% agarose gel.

Progeny Test

For the progeny and fertility test, selected sex-reversed males (genotypic females; see below) from the 32 and 34°C groups were mated individually to control brood stock females that were reared throughout at 25°C and which had proven fertility records. Eggs were collected daily and checked for fertilization and development. All other rearing procedures for the progeny test were the same as those described above.

Gene Expression Analysis

Offspring from the cross of control (XX) females with control (XY) and sex-reversed (XX) males from the 32°C group was incubated from fertilization to hatching at 25°C and 32°C, respectively. Twelve embryos were sampled from each group at developmental stages 36 and 39 (hatching) and individually analyzed for DMY/DMRT1bY and DMRT1 gene expression by reverse transcription-PCR analysis following the primers and methods of Nanda et al. [2002].

RESULTS AND DISCUSSION

Eggs that were incubated at 17 and 34°C from immediately after fertilization hatched in approximately 25 and 7 days, respectively. These results match the developmental times for these temperatures and the thermal tolerance of medaka embryos reported by Yamamoto [1975]. However, we noted that embryos incubated at high temperatures, particularly at 34°C, frequently showed abnormal embryonic development. The most common types of abnormalities were those in the nervous system such as deformed and atrophied brains, small or missing eyes, and general cranial deformities, and these developmental problems likely led to the reduced viability observed in these groups. Nevertheless, embryos with milder degrees of deformities were in general able to hatch, start exogenous feeding, and grow until adulthood, and these were the animals that were sexed and submitted to progeny testing as described below.

Determination of genotypic and phenotypic sex in sexually mature individuals from all groups was first performed non-destructively by examination of body color and secondary sexual characters, respectively. The results of these analyses were then confirmed by gene amplification and histology, respectively. Overall, the results of sex ratios compiled in table 2 clearly indicate that temperature treatment caused sex reversal from genotypic female into phenotypic male at the highest temperatures (30, 32 and 34°C). We also observed two sex reversed males at the temperature of 27°C but, since most fish in this group were accidentally lost, the actual proportion of sex reversal in this group could not be calculated. Determination of genotypic sex from body color was very accurate as all XX individuals, including sex reversed males and normal females, showed only the autosomal DMRT1 fragment by amplification with primers PG17.5 and 17.6 and none with the other three primer sets. The XX genetic makeup of sex reversed males was also confirmed in the progeny testing, as they sired all female (XX) progeny when mated to control females (results not shown). In genotypic males, however, primers PG17.5 and 17.6 amplified two fragments, a larger one (~1 kb) corresponding to DMRT1 and a smaller one (~0.7 kb) corresponding to DMY/DMRT1bY. In these animals, a single band corresponding to DMY/ DMRT1bY was also amplified with the other three primer sets.

The proportion of sex-reversal of genotypic females at the temperature of 32°C obtained in this study (40%) was grossly equivalent to that reported by Sato et al. [2005] for the same temperature (24%) if slight differences in temperature between the two experiments are taken into consideration. This study, however, shows for the first time that the efficiency of sex reversal by temperature in genotypic females of Hd-rR medaka increases proportionally with temperature above 27°C, reaching 100% at 34°C. On the contrary, our results suggest that sex differentiation in genotypic (XY) males is greatly resilient to thermal effects, as sex-reversed females were not found in any of the temperature groups. This was rather surprising, as we expected to find also sex reversal at the low temperatures. Our expectation was based on studies showing that the expression of various genes purportedly involved in the process of sex differentiation in fish is affected by environmental temperature in a similar way as by treatment with exogenous steroid hormone treatments [reviewed in Devlin and Nagahama, 2002; Strüssmann and Nakamura, 2002]. Moreover, in support of our assumption, Minamitani and Strüssmann [2003] showed that the efficiency of feminization by administration of estradiol to Hd-rR medaka embryos increased by incubating fertilized eggs at low temperature during hormone treatment. The results of this study, however, suggest that low temperature alone is not sufficient to corrupt the putative genetic mechanism of sex differentiation in medaka although we cannot rule out the possibility that exposure to even lower temperatures (for short duration to guarantee viability) could still cause sex reversal.

The above results have important implications for the discussion of the mechanism and evolution of sex determination in medaka. For example, one puzzling aspect of sex determination in this species is the paradoxical facets of the DMY/DMRT1bY gene: it is both new and old (= disappearing), and it is both sufficient but not absolutely necessary for testicular differentiation [see Schartl, 2004]. In fact, two studies [Nanda et al., 2003; Schartl, 2004] suggest that autosomal modifiers could be implicated in male fate determination in medaka. The findings of this study and those of Sato et al. [2005] add a new dimension to this discussion, as they show that part, if not all, of the masculinizing role of DMY/DMRT1bY could readily be taken over by (high) environmental temperature.

Thus, whereas Schartl [2004] has envisioned the loss of the Y chromosome of medaka and its replacement by other genetic factors, we suggest that environmental temperature may actually accelerate its demise. Indeed, it is noteworthy that sex reversal of genotypic females was observed at a temperature of 27°C in the HNI [Sato et al., 2005] and Hd-rR (this study) strains, so it is possible that at least part of the F1 and F2 XX males attributed by Nanda et al. [2003] to autosomal modifiers could have been sex reversed by the rearing temperature in their experiments (27°C). Given the occurrence of sex reversal at environmentally realistic temperatures (27-32°C) and projected trends in global warming, the postulated disappearance of the Y chromosome could be even accelerated by the 'dilution effect', in which sex reversed XX males compete with XY males for mating with XX females, leading to a gradual reduction of the proportion of XY individuals in the population [assuming the persistence of the masculinizing temperatures; see Kanaiwa and Harada, 2002]. Thus, our results point to a new paradox concerning the Y chromosome of medaka: that it is both strong, as it is able to fend off the feminizing effects of (low) temperature, but at the same time weak, as it can be readily replaced by alternative genetic and environmental factors and unable to prevent its own extinction.

A second important finding of this study was that the thermally sensitive period of sex differentiation in medaka embryos lies between developmental stages 5-6 (8 to 16 cells) and 36 (heart development stage), as thermal treatments only effectively induced sex reversal in this time/developmental window. This result conflicts with the prevailing view that gonadal sex is determined in medaka around the later half of embryogenesis (between stages 36 and 39), as implied from the results of histological analysis of gonadal sex differentiation [particularly germ cell proliferation; Satoh and Egami, 1972; Quirk and Hamilton, 1973; Yoshikawa and Oguri, 1981; Kobayashi et al., 2004].

It also contradicts estimates obtained from hormonal manipulation of sex in this species; for such treatments are 100% effective even when started after the heart development (or eyed-egg) stage [Iwamatsu, 1999; Minamitani and Strüssmann, 2003], a fact commonly taken as an indication that gonadal sex is still not determined at that time. It is worth mentioning, however, that sex steroid hormones are potent inducers of sex that can override putative processes [even in the adult gonads; Bhandari et al., 2006].

On the contrary, wherever it has been possible to determine the time of sex determination by other means, such as thermal manipulation of sex in TSD species, for example, it was shown that sex is determined considerably earlier than the first signs of gonadal sex differentiation [Conover and Fleisher, 1986; Baroiller et al., 1995; Strüssmann et al., 1997; Wang and Tsai, 2000]. For the above reasons, we consider that our results obtained by thermal manipulation reflect more closely the time of natural sex determination in (untreated) medaka than previous determinations using either histology or hormonal manipulation.

This revised estimate of the period of natural sex determination in medaka may be fundamental for the functional analysis of the role of genetic factors in this process. The DMY/DMRT1bY gene, for instance, is considered as the prime testis-determining factor in this species, with functional (but not necessarily structural) analogy to the sex determining genes in other taxa [Matsuda et al., 2002; Nanda et al., 2002]. Kobayashi et al. [2004] observed expression at stage 37, when the primordial germ cells (PGCs) have already migrated into the gonadal anlagen and before the appearance of a sexual dimorphism in the number of germ cells at stage 39. Expression of DMY/DMRT1bY was detected specifically in Sertoli-like cells surrounding PGCs, leading the authors to suggest that it regulates the differentiation of these cells and/or the proliferation/differentiation of germ cells and thereby induces testicular differentiation.

Nanda et al. [2002], on the other hand, detected DMY/DMRT1bY expression already from the neurula stage (stages 17-18), but did not elaborate on a role at this early stage. Thus, although DMY/DMRT1bY is presumably located far upstream in the male pathway, its target genes still remain unknown [Nanda et al., 2002; Schartl, 2004]. It is notable that the early window of expression observed by Nanda et al. [2002] coincides temporally with the thermally sensitive period of masculinization determined in this study, and that both occur even before migration of PGCs and their entry into the germinal ridges.

These facts suggest a similar key role for these two factors during early development, although they do not prove that the mechanism of action in both cases is the same, as for example water temperature could have cumulative effects on other genes. As expected in the absence of the gene [see also Nanda et al., 2002; Kobayashi et al., 2004], expression of DMY/DMRT1bY was not observed in any of the genotypic female (XX) embryos regardless of developmental stage and incubation temperature.

In spite of the absence of DMY/DMRT1bY, however, substantial amounts of DMRT1 mRNA were found in 6 out of 12 XX embryos incubated at 32°C, a fact not observed at the 'sexually neutral' temperature of 25°C. This finding is in good agreement with the proportion of sex-reversed males found at this temperature (e.g. 40%; table 2). This is the first demonstration that genotypic female medaka embryos express this crucial gene for testis formation in response to masculinizing temperatures. Likewise, DMRT1 expression at 32°C was observed as early as developmental stage 36. This contrasts with a temperature of 25°C, where it is observed only after hatching [stage 39; this study; Nanda et al., 2002; Kobayashi et al., 2004]. Thus, high temperature greatly accelerates the cascade of molecular events leading to testicular differentiation. Further studies must examine how high temperature triggers the expression of DMRT1, if this early expression actually induces premature testicular differentiation, and further clarify the molecular cascade leading to gonadal sex differentiation in medaka.

Histological analysis of sex-reversed males from the high temperature groups revealed that their testes were morphologically normal, presenting the characteristic tubular organization and cysts of germ cells (spermatogonia, spermatocytes and spermatozoa) [Kanamori et al., 1985], and were undistinguishable from those of males from the control group (25°C). We did not find any differences between females from the various groups either and intersex features were never observed. The progeny tests showed that sex-reversed males from the 32 and 34°C groups were capable of normal sexual courtship but only animals from the former treatment produced viable offspring. The fertility of animals from 32°C agrees with the results of hormonal [Yamamoto, 1958] and temperature treatments by Sato et al. [2005], who showed that

XX males are in general functional, although we cannot discard the possibility that they are reproductively inferior to XY males [see Fineman et al., 1975].

The infertility of sex reversed males from the 34°C treatment, on the other hand, resembles in all respects the infertile XX males observed by Yamamoto [1958] after hormonal manipulation in that they had normal testicular histology, produced mature spermatozoa, mated actively, but did not sire viable progeny. In the present study, it might represent a problem of orchestration between the neural system and the gonads since fish at the highest temperatures frequently showed cranial and brain abnormalities. This possibility and that of genotoxic effects of heat [Anitha et al., 2000] should be examined in further studies.

The results of this study show unmistakably that the process of gonadal sex differentiation in Hd-rR medaka is under the control of genetic and environmental (temperature) factors at a very early stage of embryonic development. They also reveal that a brief exposure to high temperatures early in life could greatly impair the fertility of fish as adults. Thus, medaka constitutes an invaluable research tool to study the effects of temperature on the process of sex differentiation and provides the opportunity to test the strength of a clear sex determination mechanism, which shares a high degree of homology with that of other vertebrates, in a poikilothermic context. Such studies may help to shed light on the mechanism of TSD in fish and to provide tools to forecast the impact of global warming and climatic change on the reproduction of fish populations.

CHARACTERIZATION OF THE BRAIN-PITUITARY AXIS IN PEJERREY ODONTESTHES BONARIENSIS

The pejerrey fish (Odontesthes bonariensis), a large atherinid native from the inland waters of Buenos Aires Province, is considered one of the most emblematic fresh water fish species of Argentina (López et al., 2001). Although it is an important commercial and game fish, its culture has not been well developed in Argentina. Besides, some reports have demonstrated the existence of problems in intensive pejerrey aquaculture: early sexual maturation, the presence of spawning asynchrony between females and low growth rates (Strüssmann, 1989; Strüssmann et al., 1993).

During the last years the interest of the laboratory has been focussed on the study of the reproductive physiology and the mechanisms involved during sexual determination/ differentiation in pejerrey, not only from a basic point of view but also keeping attention to the potential application of the knowledge on the development of pejerrey aquaculture.

In this context, the goal of the present work was to summarize the findings on the characterization of the brain-pituitary-gonadal axis in Odontesthes bonariensis.

Gonadotropin-releasing Hormone

Gonadotropin-releasing hormone (GnRH) is a key neurohormone that regulates reproduction in all groups of vertebrates. This decapeptide is synthesized by neurons in different regions of the brain being its best known function the regulation of gonadotropins production and release by the pituitary gland (Seeburg et al., 1987).

Today it is well known that the brain of vertebrates expresses at least two GnRH variants. However, in teleosts fish there is a growing number of species in which three different GnRH forms have been identified (Somoza et al., 2002a, b; Lethimonier et al., 2004). Up to date 14 different GnRH forms have been described in vertebrates and they are usually called using the common name of the species in which they were first described (Adams et al., 2002; Lethimonier et al., 2004). In those bony fish species expressing three variants in their brain, these GnRH forms are distributed over specific brain areas: chicken GnRH-II (cGnRH-II) is expressed by neurons of the midbrain tegmentun (MT), salmon GnRH (sGnRH) is localized in the terminal nerve ganglion (TNG) and the third species-specific form is mainly found in the anterior preoptic area (POA) and the hypothalamus (Fernald and White 1999; González-Martínez et al., 2001).

Also in pejerrey fish, three different GnRH variants have been found: sGnRH, cGnRH-II, and pejerrey GnRH (pjGnRH), a novel member of the GnRH family isolated for the first time in this species (Stefano et al., 1997; Montaner et al., 2001). The immunostaining of pejerrey brain sections showed that sGnRH is mainly expressed by neurons located at the TNG, pjGnRH at the nucleus preopticus periventricularis (NPP) of the POA, and cGnRH-II at the MT (Stefano et al., 2000). In addition, only pjGnRH was detected in fibers entering the anterior pituitary gland (Stefano et al., 1997, 2000; Somoza et al., 2002a). More recently the full length of the cDNAs encoding for the three GnRH precursors of pejerrey was obtained by the RACE method (Rapid Amplification of cDNA Ends).

The analysis of these sequence showed that the cDNAs of pjGnRH, cGnRH-II and sGnRH have 441, 530 and 516 bp, having an open reading frame of 297, 252 and 276 bp, respectively (Guilgur et al., 2003; GenBank accession #AY744689, #AY744687, #AY744688). It was also demonstrated the expression of different GnRH forms in other pejerrey organs such as: gonads, eye, kidney, spleen, liver, gill and olfactory epithelium suggesting novel functions for these peptides (Guilgur et al., 2003).

When GnRH precursors from pejerrey are subjected to a phylogenetic analysis they can be grouped following the scheme proposed by Vickers et al. (2004): pjGnRH in group one (GnRH I), containing GnRH variants located at the POA and related to the control of the pituitary gland; cGnRH-II from pejerrey in group two (GnRH II), located in neurons of the MT and sGnRH from pejerrey in group three (GnRH III), in neurons of the anterior brain (Fig. 1). In this framework it is also important to note the high percentages of identity observed between pjGnRH and seabrean GnRH (sbGnRH) precursors of different Acanthopterigian species, reinforcing the hypothesis that pjGnRH emerged from the gene codifying for sbGnRH (Montaner et al., 2001; Somoza et al., 2002b).

Gonadotropins

Gonadotropins (GtHs) are pituitary heterodimeric glycoproteins consisting of a common-subunit and a-subunit that confers hormonal specificity. Teleost fish GtHs are structurally related to the tetrapod folliclestimulating (FSH) and luteinizing (LH) hormones and are also important in the regulation of gametogenesis and sexual maturation (Quérat et al., 2000; Swanson et al., 2003).

In the last years, the genes codifying for both gonadotropins have been cloned and their expression studied in many bony fish species (see Weltzein et al., 2004). As a first step to study the function of GtHs in pejerrey, the full length of the cDNAs encoding for FSH and LH-subunits were obtained by RACE (Miranda et al., 2003a; 2004). The cDNA for FSH-SS and LH-SS have 466 and 558 bp with an open reading frame of 351 and 450 bp, respectively (Miranda et al., 2004). Comparing these sequences with the deduced aminoacidic sequence of other teleost species it can be inferred that pejerrey FSH-has a signal pepide of 15 amino acids (aa), and a mature peptide of 102 aa with a putative N-linked glycosylation site at residue Asn 12. Pejerrey LH-is composed by a signal peptide of 32 aa, a mature peptide of 118 aa and a putative N-linked glycosylation site at residue Asn 10. Pejerrey LH-, as well as all vertebrate species, has conserved the position of 12 cysteines and 1 putative N-linked glycosylation site (Swanson et al., 2003; Weltzein et al., 2004). Pejerrey FSH-, also contains 12 cysteines; however, the position and number of cysteines in FSH-varied among teleosts (Swanson et al., 2003).

The aminoacid sequence of pejerrey FSH-compared to other teleost fish sequences varies between 33.1-66.9% and is lower than that observed for LH-subunits (46.3-72.4%). These results follow the pattern indicating that the primary structure of LH-subunits has been better conserved than FSH-during teleosts evolution, suggesting a rapid divergence of the FSH-subunit (Quérat et al., 2000). Comparison studies also showed that pejerrey GtHs are closer to other Actinopterigian GtHs.

SEX DIFFERENTIATION

Odontesthes bonariensis has a strong thermolabile sex determination (TSD). The proportion of females gradually varies from 100% at 15-19°C to 0% at 29°C when the larvae are kept at different temperatures between the 1-5th after hatching

(Strüssmann et al., 1997). These results make pejerrey an interesting model to study the effects of the temperature on the mechanisms of sex determination and differentiation in teleost species.

In this species, the ovarian differentiation was first recognized at week 7 at 17°C and at week 4 at 24°C, meanwhile testicular differentiation began at week 7 at 24°C and at week 4 at 29°C. Blood vessels can be first observed at the gonads at week 6, 4, and 3 at 17, 24, and 29°C, respectively (Strüssmann et al., 2004). Thus, blood vessels are formed just before or concomitant with the first signs of commitment to either sex at all temperatures. These observations are consistent with the concept of extragonadal regulation of sex differentiation in this species.

During development, either FSH and LH, are expressed in the anterior pituitary gland during the process of gonadal determination/differentiation (Miranda et al., 2001a) and both GtHs show a clear correlation with the increase or decrease of the number of GnRH neurons located at the POA and the identification of immunoreactive GnRH fibers in the pituitary gland (Miranda et al., 2001a, 2003b). Also, the number of FSH, LH and GnRH neurons at the POA varies according to the incubation temperature suggesting a correlation with TSD. Taking together, these results suggested that the synthesis and release of GtHs may be regulated by GnRH before or just at the moment of sex determination showing that the hypothalamic-pituitary-gonadal axis is active during the time of sex determination in this species (Miranda et al., 20 Strüssmann et al., 2004).

It can also be hypothesized that GtHs regulate the production of sex steroids in the gonadal primordia and/or in the interrenal gland during the sexual determination/ differentiation period (Strüssman et al., 2004).

CONTROL OF PEJERREY REPRODUCTION IN CAPTIVITY

A series of studies were performed in order to optimize pejerrey reproduction in captivity, including the increase of sperm volume and the synchronization of spawning by hormonal treatments. In pejerrey, the sperm volume obtained by stripping of males either in natural conditions or captivity is very scarce (Strüssmann, 1989; Miranda et al., 2001b), being necessary several males to fertilize the eggs produced by a single female (Calvo et al., 1977).

In pejerrey males it was possible to increase sperm volume by environmental and hormonal treatments practically during all the year around without affecting the sperm concentration and motility. These treatments included: increase of the light phase of photoperiod, human chorionic gonadotropin (hCG), heterologous pituitary extracts and GnRH superactive analogues (GnRHA) injection (Miranda et al., 2001b; 2005).

In order to compare all hormonal treatments, an experiment using the doses previously demonstrated to induce a 5 time increase of expressible milt was performed. The treatment with sGnRH-A was the most effective. However, the use of hCG is recommended because of its effectiveness to stimulate pejerrey spermiation in low doses, its low cost and availability.

In the case of females, a commercial sustained release system (Ovaplant, Syndel, Vancouver, Canada) was used in order to synchronize spawning. Pejerrey females (250 g) in late vitellogenesis with no signals of final oocyte maturation, were intraperitoneally implanted either with pellets having 75 g of a sGnRH-A or blank pellets.

The first groups of eggs were obtained after 60 hours in the sGnRH-A group and one week later it was shown that 82% of sGnRH-A implanted females had spawned meanwhile none of the control implanted females spawned during that period. Although much work is necessary in order to find easier and non expensive ways to control pejerrey reproduction in captivity, these results demonstrated that is possible to use hormonal manipulation to induce reproduction in this species.

In this context, it is important to think that even though the use of endocrine agents can be use to induce reproduction in this species, the study of environmental and maintenance conditions can also be effective to synchronize reproduction in pejerrey.

HORMONAL SEX MANIPULATION AND EVIDENCE FOR FEMALE HOMOGAMETY IN CHANNEL CATFISH

The mechanism of sex determination in channel catfish Ictalurus punctatus was evaluated by hormonal and genetic methods. Aromatizable and nonaromatizable androgens, as well as an estrogen, caused feminization in fish fed steroids for 21 days after yolk-sac absorption.

The effectiveness of 60 g of ethynyltestosterone/g food decreased markedly when the experimental feeding period was shortened and was ineffective when the treatment lasted less than 12 days.

Females from all-female populations produced by treatment with sex hormones were mated with normal males resulting in nine spawns with a sex ratio different from 1:1. The sex ratios were statistically similar to 3 male:l female in five spawns, both 2:1 and 3:1 in two spawns, and 2:1 in two spawns. These data are consistent with a model for female homogametic sex determination in channel catfish and suggest that the YY equivalent genotype is viable.

The phenotypic sex of many fish species can be experimentally altered during early development to produce monosex populations. Generally, in teleosts, treatment with androgens results in masculinization and treatment with estrogens results in feminization (Hunter and Donaldson, 1983). Paradoxical feminization (production of allfemale populations by treatment with androgens) has been reported in cichlids (Hackmann, 1974; Hackmann and Reinboth, 1974; Nakamura, 1975).

Channel catfish, Ictalurus punctatus, were feminized by a wide range of concentrations of 17-ethynyltestosterone as well as 17-estradiol (Goudie et al., 1983).

Hormonal manipulation of phenotypic sex has been accomplished by injection into the egg, by dissolving the hormone in the water during the period of gonadal indifference, or by feeding hormone to young fish (Yamamoto, 1969).

Treatment with hormones often produces fish that are sterile, intersex (ovotesticular), or have abnormalities of the genital ducts that prevent spawning (Yamamoto,

1969; Johnstone et al., 1979a, b). Morphological anomalies that would prevent spawning have not been observed in feminized catfish.

Genotypic sex cannot currently be determined in channel catfish by chromosomal analysis. However, the sex genotype and the genetic sex determination model can be determined by mating sex-reversed females with normal male fish and evaluating the sex ratio of the progeny.

This study evaluates the influence of a variety of androgens and time of treatment on sex determination in channel catfish. Breeding evidence is provided that supports a female homogametic sex determination model.

METHODS

Hormone treatments. Channel catfish were spawned by traditional methods (Dupree and Huner, 1984). Hatched larvae were transferred in groups of 200 to 40-liter glass aquaria held at about 21°. Experimental diets were prepared by dissolving hormones in 95% ethanol, spraying the food with the solution to obtain the desired concentration of hormone, and airdrying to evaporate the alcohol.

Hormones were tested by feeding fish to satiation three times daily for 21 days starting after yolk-sac absorption. The first day that fish began to feed was designated as Day 1. The hormones tested were 17-ethynyltestosterone (17-ET), 17-methyltestosterone (17-MT), 1,4 17-methyltestosterone (1,4 17-MT), 5-dihydrotestosterone (DHT), 1-methyldihydrotestosterone (1-MDHT), 17-methyldihydrotestosterone (17-MDHT), 11-ketotestosterone (11-KT), 11-hydroxytestosterone (11-HT), norethisterone (NE), and NE plus 17-ET.

In a second set of experiments a diet of 60 μ g 17-ET/g food, prepared as described above, was administered at selected intervals within the first 21 days of feeding to delineate the minimum effective treatment period. Treatment intervals were from Days 1, 5, 9, 13, or 17 through Day 21, or from Days 1 through 8, 5 through 12, or 9 through 16.

Sac-fry were treated in a third experiment with 17-MT, 17-ET, or DHT on Days 1 and 2 after hatching and exposed under static conditions to 10 or 100 g/liter of hormone for 2 hr. Water flow was stopped and sufficient hormone solution was added to the tank to produce the desired concentration. After 2 hr, water was turned on and the hormone washed out of the tank.

Progeny sex ratio evaluation. Four-year-old fish from all-female populations produced by 17-ET or 17-estradiol (Goudie et al. 1983) were individually paired and spawned with normal males in wire cages placed in a pond. A spawning container was placed in each cage. Egg masses were removed and incubated in indoor hatching troughs. Spawning success from mid-May through mid-June under these conditions is normally about 60%.

Progeny from each treatment group were maintained and reared separately.

Sex was verified in 10-to 15-cm fish by dissection. Females had paired tubular ovarian structures (Grizzle and Rogers, 1976); in males, gonadal structures were either not discernible or consisted only of opaque convoluted lines on the ventral surface of the kidney.

Statistical analysis. Sex ratios of control and hormone-treated groups were evaluated by x2 analysis based on an expected 1 male:l female sex ratio.

The sex ratios of the progeny of hormone-treated females were tested for homogeneity by x2 analysis. When homogeneity was not indicated, groups were tested individually by x2 using Yates correction term with expected ratios of 1:1, 2:1, and 3:1 (Strickberger, 1985). A probability of 0.05 was considered to be significant.

RESULTS

Hormone Treatments

Every hormone fed to fry for 21 days (except 11-HT, which was ineffective) resulted in a significantly higher percentage of females than that in control populations. Those most effective at producing females were NE, NE plus 17-ET, 17-ET, 17-MT, and 1,4 17-MT; 17-MDHT, DHT, 1-MDHT, and KT were markedly less effective. Treatment of yolk-sac fry with DHT,

17-MT, or 17-ET added to the water at 10 or 100 g/liter did not alter the sex ratio.

The feminizing effect of 17-ET (at a concentration of 60 g food) decreased when fish were fed for less than 21 days and was absent when fed for less than 12 days. Sex ratios were not significantly affected in fish fed from Days 13 to 21 or from Days 17 to 21, nor by 8-day treatments at the beginning, middle, or end of the 21-day period.

Progeny Sex Ratio Evaluation

Six of 15 spawns produced by crossing hormone-treated females with normal males grouped into a statistically homogeneous set and had a sex ratio of 1:1.

The other 9 spawns were not statistically homogenous; however, when tested individually, 7 of the 9 spawns were not significantly different from a 3 male: 1 female sex ratio. Two of those 7 spawns and the remaining 2 spawns were not significantly different from a 2 male:1 female sex ratio.

DISCUSSION

Although the sexual phenotype of developing channel catfish could be altered by a variety of androgens and dosage rates, deviations from normal sex ratios were limited to feminization.

Goudie et al. (1983) suggested that paradoxical feminization in channel catfish may be caused by endogenous aromatization of dietary androgens to compounds with estrogenic properties. Conversion of testosterone by aromatase is part of the normal biosynthesis of estrogen (Fishman, 1982).

Some natural and synthetic androgens are not susceptible to aromatization due to their molecular structure. Aromatizable androgens used in the present study were 17-ET and 17-MT; while DHT, 11-KT, and 11-HT were nonaromatizable (Crim et al., 1981; Callard, 1984).

The two other forms of DHT-1-MDHT and 17-MDHT-are thought to be nonaromatizable due to their similarity to DHT and to the addition of methyl groups at positions that interfere with aromatase activity. Norethisterone, an estrogen component of contraceptive pills, has been found to inhibit the aromatase enzyme in vitro by some investigators (Osawa et al., 1982), but not by others (Brodie et al., 1983). Further, 1,4 17-MT has molecular similarities to both nonaromatizable androgens and to aromatase inhibitors (Numazawa et al., 1987).

Predominantly female populations were obtained by treatment with nonaromatizable androgens, suggesting that androgen aromatization may not be necessary for paradoxical feminization in channel catfish. However, Nakamura (1981) reported that 11-KT in the diet produced all-male populations of Tilapia mossambica, a species that exhibited paradoxical feminization when sexually undifferentiated animals were treated with aromatizable testosterone propionate in the water (Hackmann, 1974).

Furthermore, 11-KT treatment caused precocious appearance of male secondary sex characteristics, but no change in gonadal differentiation. Norethisterone was expected to have prevented the aromatization of 17-ET; however, NE when used alone had a feminizing effect.

Treatment with nonaromatizable androgens resulted in the lowest percentage females of any of the feminizing treatments. Higher doses or earlier treatment with any of the feminizing nonaromatizable androgens might produce different sex ratios; however, these androgens apparently have a feminizing effect during the first 21 days.

The minimum effective treatment period for production of all-female populations with 60 g 17-ET/g food must be close to the first 21 days of exogenous feeding. Reducing the duration of the treatment within any portion of the 21-day period reduced the percentage of females produced. This contrasts with phenocritical periods that are as short as 3 days in Hemihaplochromis multicolor (Hackmann and Reinboth,1974).

The six spawns resulting from pairing normal males with females from an all female population had a sex ratio of 1:1 and must have been produced by genotypic females that were unaffected by the steroid treatment.

Progeny from the nine spawns that had a sex ratio different from 1:1 must have resulted from breeding genotypic male-phenotypic female fish with normal males. These data are important for two reasons.

First, they demonstrate that the hormone treatment does not differentially affect survival of genotypic males, leaving only female fish to develop.

Second, this sex ratio is possible only if the genotypic sex determination model is equivalent to a homogametic female condition (XX).

The resulting populations would have been composed of all-male offspring if males were the homogametic sex. Additional support for female homogamety is available: all channel catfish progeny produced by gynogenesis were females (Goudie, 1987), and the sex ratio of triploid channel catfish produced by second polar body retention was 1:1 (Wolfers et al., 1981).

This female homogametic condition should be used only as a model because neither sex chromosomes nor sex-determining genes have been identified in channel catfish.

The 3 male: 1 female ratio of the progeny obtained from seven of nine sex-reversed females indicated that the male phenotype included both XY and YY genotypes.

The normal male genotype (XY) should be present in two-thirds of the males and the YY genotype in one-third. Therefore, the YY equivalent male genotype must be viable in channel catfish, as has been reported in other fish species (Hunter et al., 1982; Nakamura et al., 1984; Chevassus et al., 1988).

The YY genotype may not be as hardy as the normal XY genotype in all catfish families, which may explain why two of the populations were statistically similar to only a 2:1 sex ratio. In mammals, some nonsex-related genes essential for survival are located on the X chromosome; consequently, the YY genotype is lethal. Variability in YY viability has been found among families of medaka Oryzias latipes. The YY genotype is rare in families homozygous for xanthic coloration, but occurs with predictable frequency in families heterozygous for this trait (Yamamoto, 1975).

Identification of the YY equivalent males in channel catfish can now be accomplished only by progeny testing. All-male progeny should be produced by the spawning of YY males with normal (XX) females or with sex-reversed (XY) females.

In the first case, all offspring would have the XY genotype, and in the latter case, half of the progeny would be expected to have the XY genotype and half would be YY males.

These techniques could be useful to the aquaculture industry because male channel catfish grow faster than females (Beaver et al., 1966; Brooks et al., 1982). All-male populations could be produced without sex steroids being administered to fish reared for food.

GENETIC BIOTECHNOLOGIES IN AQUACULTURE

Genetic biotechnologies in aquaculture focus primarily on increasing growth rate, but also include improving disease resistance and increasing environmental tolerance. There are several biotechnologies that can be readily applied to farmed aquatic species.

Selective breeding, i.e. traditional animal breeding, started with the common carp several thousand years ago. However, it has only recently been applied to a handful of other species of food fish such as catfish, trout and tilapia.

As a result, many farmed aquatic species are still very similar to their wild relatives. Selective breeding programmes have however yielded significant and consistent gains; for instance a 5-20% improvement in growth rate per generation in farmed species such as Atlantic salmon, catfish and tilapia.

Chromosome-set manipulation is a technique that can be used to produce so-called 'triploid' organisms that have three sets of chromosomes instead of the usual two.

Triploids generally can not reproduce and so it was initially thought that the energy that was not channeled into reproduction would go instead to increasing growth rate, but this has not in fact proven to be the case.

The advantage of triploids seems to be rather in their functional sterility (although that does not reach 100%). For example, triploid oysters do not produce gonads and are therefore marketable at times of the year when mature oysters have an off-taste because of gamete production. ('gametes' are sex cells-the ovum, or egg in the female and the sperm in the male).

In aquaculture, one sex is often more desirable for the market than the other. For example, female sturgeon produce caviar, male tilapia grow faster than females, female trout and salmon generally grow faster than males.

The production of single sex groups of fish takes advantage of this sexual dimorphism and can be accomplished by manipulation of the developing gametes and embryo. The manipulation can be in the form of DNA denaturisation in the gametes followed by chromosome-set manipulation or alternatively, by hormonal sex-reversal and subsequent breeding.

Administering appropriate hormones can change the phenotypic (physical) sex of many aquatic species. For example, genetically male tilapia can be turned into 'physical' females through estrogen treatments.

These genetic males, that are physically female, are then mated with normal males to produce a group of all-male tilapia that grow faster and have less unwanted matings than a group of mixed-sex tilapia. (In mixed sex culture of tilapia, mating at a small size often leads to overcrowding and stunting).

Some of the all-male offspring of this process have two male chromosomes and these fish can be used in turn as broodstock for subsequent generations. A big advantage of this technique is that all-male populations can be produced in those subsequent generations without the use of any hormones.

Hybridization is another simple genetic technology that has become easier with the development of artificial breeding techniques, such as the use of pituitary gland extract and other hormones to initiate gamete development and induce spawning (the release of fish eggs).

An increased understanding of environmental cues that influence reproduction, such as day length, temperature or water current has also played an important part in improving breeding programmes.

Fish farmers can now overcome many of the natural reproductive isolating mechanisms that species develop in the wild. Hybridization can also be used to produce single sex groups of fish when the sex-determining mechanisms in the parental lines are different (for example, hybridization of Nile tilapia and the blue tilapia).

These improvements in reproductive technologies have also assisted aquaculturists greatly in their efforts to domesticate aquatic species. In addition, by making it possible to remove the natural constraints and timing of breeding, farmers are able to mate many more species at the times that are most beneficial, and thus help to ensure a steady and consistent supply of fish to the market.

Genetic engineering is a vague term that has come to be nearly synonymous with gene transfers i.e. the production of transgenic fish or genetically modified organisms (GMOs). This technology is progressing rapidly and it is now possible to move genes between distantly related species; for instance, a gene that produces an anti-freeze protein has been transferred from the winter flounder to strawberries.

Gene transfer in fish has usually involved genes that produce growth hormone and has been shown to increase growth rate dramatically in carp, catfish, salmon, tilapia, mudloach and trout. The same anti-freeze gene that was put into plants was put into salmon in the hopes of extending the farming range of the fish.

The gene did not produce enough of the protein to extend the salmon's range into colder waters, but it did allow the salmon to continue growing during cold months when non-transgenic salmon would not grow. Transgenic technology is currently in the research and development stage; to our knowledge there are no transgenic aquatic plants or animals available to the consumer.

THE SEX-SPECIFIC MARKER GENES

The model teleost medaka (Oryzias latipes, d-rR.YHNI strain) was used to produce offspring of a defined sex (monosex populations) by crossing experimentally produced YY and XX males to normal females. These monosex populations had the predicted chromosomal constitution as shown by a sex chromosome-specific DNA sequence.

However, in XX populations the spontaneous development of males without previous exposure to androgens was observed. Differences in the percentage of male offspring from individual XX breeding pairs indicate a possible variation of unknown genetic factors to be responsible for the development of XX males.

The expression of two gonadal genes that are involved in sex differentiation, Dmrt1b(Y) and Fig1a (factor in the germ line), was analyzed in monosex populations. Dmrt1b(Y) expression correlated strictly with the genotype but not the sexual phenotype. When XY juvenile fish were exposed to 17-ethynylestradiol at concentrations that induce sex reversal, Dmrt1b(Y) expression was not repressed.

However, Dmrt1b(Y) was expressed in XY or YY gonads regardless of the sex and could not be detected in XX individuals. In contrast, the expression of Fig1a correlated with the phenotypic sex: Fig1a was expressed in male juvenile fish exposed to 17-ethynylestradiol and repressed in fish exposed to 17-methyltestosterone. The Dmrt1b(Y) expression appears to reflect an early and important event in sex determination and lends support to the suggested key regulatory role of the Dmrt1b(Y) gene in sex determination. This process is apparently hormone insensitive, and the expression of further downstream acting genes can be regulated (directly or indirectly) by sex steroids.

In mammals, the sex determining gene Sry located on the Y chromosome triggers the development of the male phenotype. A cascade of potential downstream genes including Dmrt1, Sox9, Amh, and others is involved in early male-specific differentiation. Although in other vertebrates with genetic sex determination, no homolog of the Sry gene has been identified, genes acting downstream of the mammalian Sry show a conserved expression pattern in many vertebrates, regardless of whether the sex is genetically or environmentally determined. For example, the transcription factor Sox9 and the gene for the anti-Muellerian-hormone are involved in the early differentiation of the male phenotype in both mammals and reptiles, and Dmrt1 is expressed during testicular differentiation in mouse, chicken, and rainbow trout. Recently a different sex-determining gene of the medaka (Oryzias latipes) has been identified. The gene Dmrt1b(Y) (for Doublesex-and Mab-3-related transcription factors), a duplicated copy of the autosomal Dmrt1, is the only functional gene in the Y specific region and is expressed in pre-Sertoli and Sertoli cells (the gene was originally called DmY or Dmrt1Y; according to the nomenclature proposed by Volff et al., Dmrt1b(Y) will be used throughout the text). The expression pattern makes it very likely that Dmrt1b(Y) is the male sex-determining gene, but the biochemical function of Dmrt1b(Y) is not known. The DM-domain, a DNA-binding motif, is also found in Drosophila doublesex (Dsx) and Chaenorabditis Mab-3 genes, which are both expressed downstream in the sex-determination cascade.

In contrast to mammals, sex determination in many fish including the medaka is labile, and the genetic sex can be reversed functionally by exposure to androgens or estrogens during a certain period of early juvenile development. Monosex populations, offspring with only one sex, can be produced by crossing experimentally produced YY or XX males with normal females. This approach has been widely used in aquaculture because one sex is often favored for fish production because of sex-specific growth differences. For scientific purpose, monosex populations allow the study of sex-specific genes during early development when phenotypic sex differences are not yet discernible. In the present study, we report the application of an improved protocol for the establishment of monosex populations using the d-rR.YHNI strain of the model teleost medaka. This strain is characterized by several sex markers: 1) The phenotypic sex can be determined by sex-specific fin characters (shape and size of the anal fin); 2) the genetic sex is visible by the orange body color because the R-locus coding for a gene involved in the formation of orange chromatophores is active only on the Y chromosome; and 3) a deletion in the SL1-locus on the Y chromosome allows identification of the genetic sex in embryos. Amplification with primers flanking this deletion produces polymerase chain reaction (PCR) products of different length for X and Y chromosomes and allows the XY and YY genotypes to be distinguished. Both the r-and SL1-loci are sex-linked markers in the pseudoautosomal region and are not sex determining. Their distance to the sex-determining locus is 1.6-2.1 cM for SL1 and 0.36 cM for the r-locus.

The established monosex populations were used to study the effect of sex hormone exposure on two early expressed sex-specific genes: Dmrt1b(Y), the sex-determining gene of medaka and (factor in the germline), an oocyte-specific gene expressed early in development.

MATERIALS AND METHODS

Fish Culture

The medaka d-rR and Hd-rR.YHNI strains were kindly provided by Dr. Y. Wakamatsu (laboratory of fresh water fish stocks, Bioscience Center, Nagoya University, Nagoya, Japan) and Prof. M. Sakaizumi (Department of Environmental Science, Niigata University, Niigata, Japan). The animals were kept under a constant cycle of 16L:8D at 26°C (±1°C). Fish were fed ad libitum three times daily, once with Artemia and twice with commercial flake food (TetraMin, Tetra GmbH, Melle, Germany). Breeding was performed in 10-L tanks containing one male and one to three females.

Exposure of Fish

Exposure concentrations and conditions that are known to induce complete functional sex reversal were chosen. All exposure schemes started at the time of hatching. Genetic males were exposed to 17-ethynylestradiol (EE2, purity >97%, Merck, Darmstadt, Germany) added to the water. Because addition of 17-methyltestosterone (MT, purity >98%, Sigma, Deisenhofen, Germany) to the water caused sterility in all sex-reversed XX males (unpublished data), fish were fed with MT-containing flake food.

Additional feeding of fish with Artemia may lead to a reduced uptake of contaminated flake food. Thus, to ensure sufficient dosage of MT for sex reversal and according to the original protocol of Yamamoto, fish exposed to MT (and appropriate control fish) were not fed with artemia.

EE2 was dissolved in dimethylsulfoxide (DMSO, purity 99%, Sigma) and added to the aquarium water to reach a final concentration of 100 ng/L EE2 and 0.005% (v/v) DMSO. Control fish were exposed to the same DMSO concentration. To avoid metabolic and microbial breakdown of EE2, 90% of the water was removed and replaced every 2 days with fresh EE2-contaminated water. Every 7 days all groups of experimental animals were placed into new aquaria with completely exchanged tank water.

Sex-reversed fish were produced by exposing them for 2 months to EE2. Following the exposure the animals were kept in clean water until they had reached sexual maturity. To prepare MT-containing food, 25 μg MT was dissolved in 5 ml ethanol, mixed with 1 g flake food (Tetramin, Tetra), and the ethanol was evaporated overnight. Fertile sex-reversed XX males were produced by exposure for 3 months. Following the exposure the animals were fed with normal food until sexual maturity.

Preparation of the d-rR.YHNI Strain

Males of the Hd-rR.YHNI strain were crossed with females of the d-rR strain. Males from the offspring of this strain were crossed once again with females of the d-rR strain. The resulting strain was called d-rR.YHNI and maintained as a stock of approximately 100 individuals per generation. Animals used for the experiments were from the fourth-generation offspring.

Production of Monosex Populations

Monosex populations were produced according to Yamamoto with some modifications. Offspring of the d-rR.YHNI strain were exposed to EE2 to produce XY females. XY females were identified by female fin characters associated with the Y-linked orange body color and crossed with normal XY males. Twenty-five percent of the offspring could be expected to be YY males, which could not be distinguished from XY males by fin morphology or body color. By analyzing the sex chromosome-specific DNA sequences, YY males could be identified and were crossed with XY females. The offspring of this cross was exposed to EE2 and YY females were identified by analysis of sex chromosome specific sequences. To maintain a stock of YY medaka, YY males and YY females were mated and some of their offspring were exposed to EE2. XX males were obtained by crossing normal d-rR.YHNI males and females and treating the offspring with MT. XX males could be identified by their body color. To maintain a stock of XX medaka, XX males were crossed with XX females and some of their offspring were exposed to MT.

PCR of Sex Chromosome-Specific Sequences

DNA was isolated from either a 2 x 3 mm^2 piece of the anal fin or whole 6-day-old embryos after removing the chorion, according to Shinomiya et al. Briefly, fins or embryos were incubated for 60 min in 100 μl lysis buffer (0.1 mg/ml proteinase K, 100 mM EDTA, 50 mM Tris pH 8.0, 100 mM NaCl, 1% SDS). Then 100 μl of chloroform:phenol (1:1, buffer saturated) was added and mixed by vigorous shaking for 1 min. After centrifugation for 5 min at 12 000 g, DNA was precipitated from the upper phase by adding one tenth of the volume sodium acetate (3 M, pH 5.2) and an equal amount of isopropyl alcohol. After cooling for 20 min at-20°C, the samples were centrifuged for 10 min at 12 000 x g, washed with 70% ethanol, and air dried. The DNA pellet was solubilized in 50 μl of sterile water. PCR was performed using the primers pHO5.5F (5'-CCT GCA ATG GGA AAT TAT TCT GCT C-3') and pHO5.5RV (5'-CTT TTG TGT CTT TGG TTA TGA AAC GAT G-3') at 55°C annealing temperature and 30 cycles.

Reverse Transcription-Polymerase Chain Reaction

RNA was prepared from medaka embryos or juvenile fish with Trifast (Peqlab, Erlangen, Germany) according to the manufacturer's instructions. If juvenile fish were used, head and tail were removed prior to RNA analysis. Because of the different exposure conditions, growth of juvenile fish exposed to MT was delayed when compared with fish exposed to EE2. To analyze gene expression in juvenile fish of the same size and stage of development, samples for RNA isolation were taken at 23 days for fish exposed to EE2 and 40 days for fish exposed to MT. The cDNA was synthesized from DNAse (Roche Diagnostics, Mannheim, Germany)-treated RNA using oligo dT primer and Muloney murine leukemia virus reverse transcriptase (Invitrogen, Karlsruhe, Germany). PCR was performed at 60°C annealing temperature for 30 (Actin, Figla) or 40 cycles (Dmrt1b(Y)) using Taq-polymerase (Peqlab) with 0.5 µl cDNA per 15 µl reaction volume. Actin transcripts were amplified using the primers 5'-AGG GAG AAG ATG ACC CAG ATC-3' and 5'-CGC AGG ACG CCA TAC CAA-3'. Dmrt1b(Y) was amplified using 5'-CCG GGT GCC CAA GTG CTC CCG CTG-3' as forward and 5'-GAT CGT CCC TCC ACA GAG AAG AGA-3' as reverse primer. Figla was amplified using 5'-TAC TGC TGC ATC GAG AAG TAC AAG-3' as forward and 5'-AGA GTA CAG CTG AAA GCA AAA TGA G-3' as reverse primer. The primers for Actin and Dmrt1b(Y) were cDNA specific (i.e., because of introns in the sequence flanked by primers, amplification of genomic DNA resulted in larger fragments.)

Alternative primers were used for the amplification of Dmrt1b(Y) and actin in embryos: For the detection of Dmrt1b(Y), the primers 5'-GGC CGG GTC CCC GGG TG-3' and 5'-TGC GGC AGA CAG AGG ATT GG-3' were used. Actin was amplified with 5'-TTA AAC AGC CCT GCC ATG TA-3' as forward and 5'-GCA GCT CAT AGC TCT TCT CCA GGG AG-3' as reverse primers.

Statistical Analysis

Statistically significant differences of the proportions of male offspring ($P < 0.05$) were analyzed by the Fisher's exact test using the program GraphPad InStat, version 3.00 (GraphPad Software, San Diego CA, www.graphpad.com).

RESULTS

Production and Characterization of Monosex Populations

Matsuda et al. have established the medaka inbred strain Hd-rR.YHNI with a sex chromosome-specific DNA sequence. This strain showed reduced fecundity in our lab (i.e., the first spawning of females was delayed 1-2 months and fewer eggs were produced when compared with the d-rR strain). The d-rR strain [12] has the same Y-linked orange color marker as the Hdr-R.YHNI strain but lacks the sex chromosome-specific DNA sequences. By crossing of d-rR females with Hd-rR.YHNI males, the sex-specific chromosome sequence was introduced into the offspring of these parents. The resulting strain was called d-rR.YHNI, and no difference in fecundity could be observed, compared with the d-rR strain.

By means of a series of hormone exposures and crossings described by Yamamoto, the d-rR.YHNI strain was used to produce XY females, YY males, YY females, and XX males. Because YY medaka cannot be distinguished from XY medaka by morphological criteria, Yamamoto performed backcrossings to identify the YY genotype. In contrast to the d-rR strain used by Yamamoto, the d-rR.YHNI strain allowed YY males and females to be identified by analysis of sex chromosome-specific DNA sequences. PCR of DNA from XY fish resulted in two distinct bands, but XX and YY fish showed only one specific band corresponding to either the X or Y chromosome. By isolating DNA from the anal fin, YY animals were identified by the lack of the X-specific fragment after amplification of sex chromosome-specific DNA. The same fish from which the DNA samples of fin tissue had been taken were used in subsequent mating experiments. The genetic sex of the offspring was analyzed by isolating DNA from individual 6-day-old embryos. The determined genotypes of the offspring were consistent with the known genotypes of the parents:

(1) a cross of normal XY males and XX females yielded embryos of either the genotype XY or XX;

(2) offspring of the cross of YY males and XX females all had the male genotype;

(3) by crossing sex-reversed XX males with XX females, 100% of the offspring were of the XX genotype; and

(4) the offspring of YY males mated with sex-reversed YY females were all of the YY genotype.

To test whether the phenotype matched the genotype in all crosses, the offspring were reared until sexual maturity and the phenotypic sex was determined by observation of the anal fins and dissection of the gonads. In keeping with the Y-linked orange body color, all orange fish were phenotypic males. However, some males (2%) with white body color developed from the offspring of XY males mated with XX females. Among the offspring of sex-reversed XX males with XX females, 17% developed as males. Analysis of the sex chromosome-specific DNA confirmed the XX genotype of these males. Furthermore, the lack of Dmrt1b(Y) expression indicated that male differentiation was not caused by Dmrt1b(Y) transferred to the X chromosome by a historical cross-over event. These result confirmed the findings on spontaneous sex-reversed XX males obtained for other strains of the medaka.

The reason for the occurrence of male differentiation in XX individuals is unclear. To test whether we could select for (or against) spontaneous male differentiation in this genotype, we analyzed offspring of 10 different XX mating pairs. Indeed, pairs with no male offspring and pairs with male offspring varying between 3.6% and 18.5% were observed. Statistical significance was reached only when the mating pair with the highest proportion of males was compared with pairs with no male offspring or compared with all other pairs. By testing the presumption that pairs with a certain percentage of male offspring belong to similar groups of a certain genetic predisposition, a statistically significant difference could be found for pairs with a high proportion of male offspring when compared with pairs with a low proportion of male offspring.

If the difference in the percentage of XX males between individual mating pairs was based on a genetic difference, mating of spontaneous XX males (i.e., males that have occurred without exposure to hormones) to normal females might result in an increased percentage of male offspring in the following generation. However, this increase was not observed.

Expression of Sex-Specific Genes in Monosex Populations

The XX mating pairs that previously did not show any spontaneous development of XX males in their offspring were chosen to analyze gene expression and its relationship to phenotypic and genotypic sex.

Using the previously established monosex populations, Dmrt1b(Y) expression was detected specifically by RT-PCR in RNA pools of XY embryos. However, Dmrt1b(Y) was not expressed in XX embryo populations.

To analyze the effect of hormonal sex reversal on the expression of Dmrt1b(Y), fish were exposed to EE2 or MT. Because Fig1a could not be detected in female (XX) fish prior to 23 days after hatching, the reverse transcription (RT)-PCR analysis of gene expression after sex reversal was carried out at this stage of development. Fig1a was expressed in normal juvenile XX females and experimentally produced sex-reversed XY females, but XY males did not express this gene. The lack of expression of Fig1a in one sex-reversed XY female may be attributed to a delay in growth, which has been observed occasionally for individual fish. In contrast to Fig1a, expression of the master regulating gene Dmrt1b(Y) appears to be linked to the genetic sex: Dmrt1b(Y) expression was not repressed in sex-reversed XY females. Because Dmrt1b(Y) is located only on the Y chromosome, no expression was observed in XX medaka. Some very weak bands in RT-PCR samples of XX females in the size of the Dmrt1b(Y) band were unspecific artifacts because Dmrt1b(Y) principally cannot be detected in genomic DNA preparations of XX medaka (data not shown).

DISCUSSION

Analysis of sex specifically expressed genes in early development of fish requires sexing of individual embryos or larvae. For the model teleost medaka (Oryzias latipes), several markers for the early identification of the genetic sex are available if appropriate strains are used. One marker is the d-rR strain, which allows the identification of genetic males by observation of orange chromatophores (r-locus). However, orange chromatophores cannot be identified prior to approximately 2 wk after hatching, and the orange color is often weak and difficult to observe. Furthermore, in a small percentage of cases (<0.5%) the color marker does not segregate with the male sex and orange females develop. In these exceptional cases, presumably cross-over between the r-locus and a sex-determining locus on the Y chromosome took place. Another marker is in the Quart strain, in which formation of leucophores (lf-locus), visible in the embryo from 4 days post fertilization, is restricted to males. Similar to the r-locus, an inferred recombination frequency of 2.2% was observed. Yet another marker, the Hd-rR.YHNI or d-rR.YHNI strains (this study), offers the possibility of identifying the sex of embryos by analysis of sex chromosome-specific DNA. However, this technique would require the simultaneous isolation of DNA and RNA in individual fish.

The production of monosex populations is an alternative approach circumventing the sexing of individual fish. The present study has demonstrated that monosex population can be established and controlled using a molecular marker. YY and XX males were mated with normal females, and the unique genetic sex of the offspring was detected by the analysis of sex chromosome-specific DNA sequences. For male monosex populations, genetic sex clearly matched the phenotypic sex. However, XX monosex populations consisting of only female genotypes had up to 18.5% phenotypic males. Because these males were lacking the expression of the sex-determining gene Dmrt1b(Y), it could not be excluded that spontaneous development of XX males without exposure to androgens is caused by a combination of unknown genetic factors. Different frequencies of XX males in medaka, depending on the strain, have also been observed: Some strains did not show any occurrence of XX males, suggesting that the development of XX males might be caused by unknown genetic factors in the d-rR.YHNI strain. Crossing the offspring of those mating pairs with a high percentage of XX males (this study) did not result in a further increase in the percentage of XX males or at least a reproducibly high percentage.

The genetic (or epigenetic) basis for the observed instability remains to be elucidated. To establish XX female monosex populations with virtually no male offspring, selection of appropriate XX breeding pairs will be continued for several generations and may support or disprove a genetic basis for the spontaneous development of XX males.

Monosex populations were used to analyze Dmrt1b(Y) and Fig1a, two sex-specific gonadal genes expressed during early development. Dmrt1b(Y) could be detected in 7-day-old embryos of male monosex populations but was not detectable in XX embryos. Dmrt1b(Y) expression was hormone insensitive and appeared to be linked to the genetic sex: If juvenile XY fish were exposed to EE2 at sex-reversing concentrations, expression of Dmrt1b(Y) was not repressed. In contrast, Fig1a was hormone sensitive, could be induced in genetic males by exposure to EE2, and was repressed by MT.

The difference in the expression of both genes is in keeping with their suggested function. Although not structurally related, Dmrt1b(Y) resembles the mammalian Sry master regulating function for sex determination. It is probably the first gene in the male differentiation pathway of medaka, and it may not have any function other than initiating a gene cascade leading to male development.

The apparent hormone insensitivity of Dmrt1b(Y) supports this key regulatory function in sex determination. Unlike Dmrt1b(Y), Fig1a is not a master regulating gene but, with respect to the sex-determining cascade, is probably a further downstream acting gene involved in the phenotypic differentiation of the ovary. The gene is expressed during early oocyte development and has been identified by subtractive cloning in fish larvae just 1 day after hatching. The mouse homolog is a germ cell-specific transcription factor involved in the co-ordinate expression of the zona pellucida genes. Its linkage to an important structural component of the oocyte explains its expression in genetic males exposed to sex-reversing estrogen concentrations.

Analysis of the sex specifically expressed genes Dmrt1b(Y) in medaka and the effect of hormone exposure have clearly demonstrated the usefulness of monosex populations. The relatively easy experimental manipulation of sexual differentiation in fish and the analysis of key genes present in all vertebrates offers an attractive model system to unravel the molecular mechanisms of sex determination and differentiation.

AQUACULTURE AND MARINE BIOTECHNOLOGY

Main Thrust of Activities

- Development of recombinant diagnostics and vaccines for major diseases in aquaculture
- Molecular biology of Indian species, identification of useful genes and development of indigenous gene constructs for fish transgenesis work
- Genomics & proteomics studies in aquaculture species
- Bioactive molecules from aquatic organisms for therapeutic and industrial applications
- In vitro tissue culture, cell culture system development in various aquaculture species
- Front-line demonstrations to prove techno-economic viability of seed production and aquaculture in non-traditional species
- Programmes on new feed development, nutrition, breeding, spawning agents, health and post harvest aspects etc. for enhancing aquaculture productivity
- Programme support in marine biotechnology
- Training Programme in Molecular Biology for Fisheries Scientists

Achievements

Diagnostics and vaccine development in aquaculture are a major priority apart from development of suitable cell culture systems, immunostimulants, anti-bacterial and anti-cancerous agents. Studies have been supported on isolation of novel enzymes from deep-sea fungi, molecular cloning and gene expression for alkaline protease and molecular epidemiology of white spot shrimp disease in brackish water system. DNA vaccines developed against Aeromonas are being studied for the efficacy and challenge experiments. Immunodot-based diagnostic kit is being used for detection of white spot syndrome virus (WSSV) in shrimp and prawn. An ELISA-based kit has been developed for detection of IgM of Indian major carps against Aeromonas hydrophila infection. An anti-WSSV formulation was made as a prophylactic measure in shrimp health management.

Biosurfactants production was standardized from marine wastes employing the acinetobacter strains. New source of molecular fluorescent dyes has been attempted from sea cucumber tissues for bioactive compounds and use in biomedical applications. Bioactive properties of ink gland of cuttlefish have been studied for antibacterial, antiviral and anticancer agents. Cell culture attempted on brackish water species, seabass, could lead to establish stable species-specific cell lines useful in virus isolation. The growth-promoting factor studied on development of embryos of Indian horseshoe crab showed significant enlargement of organ development. Development of culture technology in non-conventional species like seaweeds and product and process development of commercial importance was also the part of the programme. A protocol on better aquaculture management practices was developed through recirculatory and raceway system. In service training programme in molecular biology for fisheries scientists was received greater attention in the programme funded at CIFT, CIFA, CIFE. Two programme supports on Aquaculture/Marine Biotechnology have been supported.

Diagnostics & Vaccines Development

At College of Fisheries, Mangalore, work was carried out on the cloning and expression of virulence genes of Aeromonas hydrophila, as an important fish pathogen and the attempts were made to isolate the proteins and study the immunogenicity. At Centre for Biotechnology, Jawaharlal Nehru University, New Delhi, recombinant porin has been expressed and purified in large quantity for maintaining the consistency of the porin vaccine.

The efficacy study and challenge experiments are being undertaken on immunization of mice and fish with porin DNA vaccine. Studies were undertaken on pathogenecity of the Epizootic Ulcerative Syndrome (EUS) in fish at Viswa Bharti University, Santiniketan; North Bengal University, Darjeeling and Center for Cellular and Molecular Biology (CCMB), Hyderabad.

A project was undertaken at College of Fisheries, Mangalore on expansion of panel of monoclonal antibodies and standardization of immunodot test kit for detection of white spot virus of shrimp. A simple monoclonal antibody (MAb)-based Immunodot kit for detection of WSV of shrimp at the farmers level has been developed, which is a cheap and effective alternative for expensive and complex DNA based tests. Development of DNA-and protein-based diagnostic methods of white spot disease of shrimp, their field evaluation and monitoring were undertaken at Anna University. The purified viral protein was used as diagnostic assay kit and transferred to Posideon Aquatech Limited, Chennai for commercialization. An ELISA test has been developed using rabbit antisera against carp IgM for the detection of IgM in IMC.

Studies were undertaken on development of anti-WSSV formulation against anti-white spot syndrome virus compound as a prophylactic measure in shrimp health management at CAH Abdul Hakeem College, Melvisharam. A process to purify the compound responsible for inactivating the WSSV was optimized.

Cell Culture in Aquaculture

Development and characterization of cell lines from selected finfish and shell fish was undertaken at CIFE, Mumbai and a cell culture system was developed from the fish Lates calcarifer (Sea bass). At CUSAT, the cell lines developed from eyestalk sub-cultured and made available to develop such cell lines as and when required.

The cells shown differentiation in to a proliferating bunch of cells and subsequently got detached as a lump of cell aggregates. The development of cell lines from seabass, Lates calcarifer was attempted at CAH Abdul Hakeem College, Melvisharam and CUSAT. The kidney cells shown fibroblastic and formed a monolayer. The cells were were deposited at NCCS.

Growth Promoting Factor(s)/Signal Transduction

A collaborative project was undertaken on identification, isolation and characterization of growth promoting and differentiation factor(s) from perivitelline fluid of developing embryos of the Indian horseshoe crab at Agharkar Research Institute, Pune, National Centre for Cell Science (NCCS), Pune and National Institute of Oceanography (NIO), Goa. Active fraction identified shown significant enlargement and enhanced compartmentalization of the heart in cardiac development and promoting activity.

In another project on large-scale cultivation of amoebocytes in vitro from Indian horseshoe crab implemented by NCCS and NIO, regeneration of gill lamellae of the Indian horseshoe crab was successful in amoebocyte production. A molecular signal intervention mechanism was developed as an alternative to eyestalk ablation at the Departments of Zoology and Biochemistry, University of Madras. A specific inhibitor that intervenes in this pathway has been shown to reduce the molt cycle of the prawn and reduces the turnover time and rearing cost and maximizes the economic returns. By decreasing the molt cycle duration, it can accelerate the growth rate of the prawn resulting in early maturation.

Feed Supplementation for Prawn

Studies on feed supplementation of highly unsaturated fatty acids (HUFAs) and immunostimulants for disease resistance in freshwater prawn was carried out at the Department of Aquaculture, College of Fisheries, Mangalore. HUFAs namely Eicosapentaenoic acid (C20: 5n-3) and Docosahexaenoic acid (C22: 6n-3) are important for larval growth. The supplementation of PUFA incorporated feed could enhance growth, survival and disease resistance in freshwater prawn.

Raceway based Shrimp Farming

A project on prototyping of raceway based third generation shrimp production technology was implemented at the Fisheries Biotechnology Centre, Fisheries College and Research Institute, Tuticorin. The protocol has been developed to increase survival of post-larvae. This in-door shrimp farming facility will be eco-friendly, economically viable and technically useful for large-scale adoption in aquafarming systems in India.

Artemia Production

The work on development of Artemia franciscana culture in the solar salt ponds and processing of Artemia cyst and biomass was undertaken at Centre for Marine Science and Technology, Manonmaniam Sundaranar University, Rajakamangalam, Tamil Nadu. Artemia production was initiated in condenser ponds and inoculated in culture ponds and the efforts are being made to utilize the abandoned area of the saltworks for Artemia cultivation.

Product/Process Development

Studies were undertaken at NIO, Goa to isolate, purify & characterize new natural fluorescent dyes and fluorescent molecules from tissues of sea cucumber Holothuria scabra and to test them as bioactivities compounds for therapeutic and industrial applications. At Kerala Agriculture University, Kochi, attempts were made to isolate bioactive substances such as antibacterial, antiviral and anticancer agents from the accessory nidamental gland and ink of cuttlefish (Sepia pharaonis ehrenberg). An antitumour peptidoglycan was isolated from cuttlefish ink and the crude preparation showed antitumour activity against mice following intraperitonial administration.

The peptidoglycan was made up of protein and uronic acid rich with polysaccharide fraction and was found to be made up of five amino acids. The screening of marine Acinetobacter genospecies for production of biosurfactant/s, purification and antimicrobial activity was undertaken at Pune University. Untapped natural sources were screened for novel biosurfactant/s and bioemulsifier-producing microorganisms using Acinetobacter strains, which showed emulsification of toluene, diesel, hexadecane and crude oil, haemolysis of human RBC's and anionic biosurfactant/s production.

Novel Enzymes

At the department of biotechnology-CUSAT, the work on molecular cloning was undertaken to develop a technology for production of alkaline protease enzyme using marine fungus, Engyodontium alba. The optimal conditions were studied for enzyme production using solid-state fermentation of wheat bran as the solid substrate. The enzyme was found relatively stable for one year. At NIO, Goa, fungi from deep-sea sediments were explored as new sources for protease. Fungal growth and protease production at low temperature and elevated hydrostatic pressure was studied to understand their adaptations to extremophilic conditions.

Seaweed Cultivation

At CSMCRI, Bhavnagar, the project was undertaken on scale-up cultivation and processing of phycocolloid seaweeds (Gelidiella acerosa and Eucheuma). Multilocational trials undertaken to validate different culture methods for large-scale cultivation of Eucheuma proved that the raft cultivation method consistently yields better harvest. This method has been successfully demonstrated in the Gulf of Mannar, Tamil Nadu coast as well as at Okha, Gujarat coast. High quality agar was extracted from Gelidiella acerosa and Gracilaria sp. The Gracilaria agar had very high gel strength (1200 g cm-2) with low gelling temperature. This agar will be useful in bacteriological and molecular biology applications.

Bioreactor for Nitrifying Water for Prawn Hatcheries

The nitrifying bioreactor technology developed at CUSAT paves the way for transforming the open prawn/shrimp larval rearing systems to closed re-circulation systems. The reactor, on properly activating and deploying in the hatchery system, can take care of the entire water quality problems. Apart from nitrification this duel capability lowers down the biological oxygen demand, total bacterial population along with stabilization of Vibrio population.

In-service Training Programme in Molecular Biology

The training programme for fisheries scientists was undertaken at three institutes namely Central Institute of Freshwater Aquaculture (CIFA), Bhubaneshwar, Central Institute Fisheries Technology (CIFT), Cochin and Central Institute of Fisheries Education (CIFE), Mumbai to train in service fisheries scientists on aspects of modern biology, health management and conservation of resources etc.

Trainees are being given both theoretical hands-on training and independent project work on molecular biology aspects relevant to fisheries topics. After successful completion of the training programme, the trainees have started working in the area of molecular biology in marine sciences at their institutes.

New Programme Support

The programme on aquaculture and marine biotechnology has been identified for support after a detailed discussion under the Task Force on Aquaculture and Marine Biotechnology. Two proposals from the UAS, College of Fisheries, Mangalore and the Cochin University of Science and Technology (CUSAT), Cochin have been supported for attention on brood stock and larval health management of black tiger shrimp and freshwater prawn and screening for potential anticancer agents from marine Cnidarians, bacteria & fungi.

Genotypic characterization and sequencing of genes of interest of marine bacteria, fish and shellfish immunology and screening for useful natural products from microorganisms, capacity building through hands-on training programmes for scholars and technicians on molecular techniques for detection of pathogens and health management in shrimp hatchery system, microbial taxonomy, crustacean immunology and screening and characterization of marine microbes for useful products etc. are also the part of the programme.

Shrimp Genomics

Shrimp (Penaeus monodon) is a very important species for the country from the commercial viewpoint. Unfortunately, the species is not domesticated and the harvests are mainly from the wild.

The reproductive biology is complex and the disease burden very high, especially viral diseases. There are populations, which are naturally resistant to viral attack and could be tapped to our advantage.

The network programme on shrimp genomics has been developed covering the aspects viz. functional genomics in disease resistance of Penaeus monodon, gene regulation of maturation and breeding in P. monodon, microsatellites and physical mapping of P. monodon and genomics of Penaeus indicus for comparative genomics.

Technologies/Products

(a) Development of a heat killed whole cell vibrio vaccine using a virulent strain of Vibrio harveyi was developed at College of Fisheries, Mangalore. This has shown immune response in shrimp (Penaeus monodon). The technology has been transferred to Mangalore Biotech Lab.

(b) A polyclonal antibody based immunodiagnostic kits for detection of different bacterial pathogens in finfish and shellfish has been developed by CIFE, Mumbai. The technology has been transferred to GlaxoSmithKline Pharmaceuticals, Mumbai without any modifications for commercialization. The kits are useful as on-farm, rapid diagnostic kits to detect bacterial and viral diseases of finfish and shellfish.

(c) To overcome the problem of viral diseases in shrimp aquaculture and for simultaneous detection of WSSV and MBV, highly sensitive method "Combi kit-WSSV, MBV", a test kit, for specific detection of pathogens is developed by the Department of Microbiology, UAS, College of Fisheries, Mangalore. The test kit can detect all genotypes of WSSV and can be used to detect the presence of virus in shrimp, pond water, sediment, carrier animals etc. The technology has been transferred to Mangalore Biotech Laboratory for packaging and commercialization.

(d) A polyclonal antibody-based immunodiagnostic assay for the detection of white spot syndrome virus-a simple diagnostic test kit developed in joint collaboration with the Centre for Biotechnology, Anna University and C. Abdul Hakeem College, Melvishram was transferred to an aquaculture-based company (Poseidon Biotech, Chennai)

Patent Obtained

In vitro production of amoebocytes from Tachypleus gigas in leibovitz culture medium (U patent # 6,790, 659 dated 14 September 2004) by Ramesh R. Bhonde and Anil Chatterji (2002)

Research Priorities for Funding

The support is continuing in the area of Aquaculture and Marine Biotechnology and the emphasis is on new programmes in the following priorities areas identified:

" Structural and functional genomics related to indigenous freshwater / estuarine / marine species

" Marine pharmaceuticals; expression of novel genes coding for bioactive macromolecules of therapeutic and industrial applications

" Marine biomaterials, bio-adhesives, bio-flocculent, bio-surfactants, medical implants, biopolymers and bioplastics

" Novel enzymes, biosensors

" Bioremediation/antifouling agents from aquatic organisms

" Development of diagnostics kits and recombinant vaccines for disease management in aquaculture

" Development of cell/tissue culture in cultivable marine organisms, invertebrates and seaweeds etc.

" Fish transgenics for therapeutic and ornamental purpose

" Marine extremophiles

10

USES IN DRUG AND VACCINE MANUFACTURE

INTRODUCTION

It is time for world-wide co-operation to gain approvals of drugs and licensure of biologics for aquaculture. At the Workshop on International Harmonization for Aquaculture Drugs and Biologics held in February 1997, a start was made toward that goal. One of the five committees set up to help this process was the one to identify priority drugs and vaccines. A Workshop and Round table were held on this topic at the EAFP Eighth International Conference on Diseases of Fish and Shellfish, Edinburgh, Scotland, September 14-19, 1997. The Workshop, "Models of Partnership for Registration of Drugs and Vaccines", and the Round table, "Progress with Registration of Drugs and Vaccines for Aquaculture", identified

(a) what aquaculture drugs are approved and what vaccines are licensed in each country or continent;

(b) the top priority drug and vaccine needs for approval and licensure in each country or continent;

(c) certain guidelines and regulations for approval or licensure;

(d) the researchers and the institutions that are performing the needed research or that have the capability and expertise to perform this research; and

(e) existing partnerships throughout the world that have worked together to gain legal products for aquaculture.

This Workshop and Round table did not completely identify the standards and guidelines required in each country but a determination was made on which ones are the most important to harmonize. Pharmaceutical, chemical, and vaccine companies provided great insight into this area. The Workshop and Round table helped the communications process by setting up mechanisms to keep all countries informed of the progress in this area. This communication network then can help to educate persons about the processes and to advocate for co-operation.

Aquaculture Drug Approvals in Each Country or Continent

The drug approval process varies in each country and continent. A brief summary follows for each continent along with tables that list the approved drugs and licensed vaccines. These tables do not indicate the diseases and species for which they are indicated.

AFRICA--Some of the 50 countries in Africa have aquaculture industries but most countries do not have any regulations regarding the use of drugs or vaccines.

ASIA--The control of aquaculture drugs in the region varies among countries from no regulation to restrictive regulations. In countries where regulation is in effect, the process for approval of drug for food animals including those used in aquaculture is complicated, time consuming, and expensive. Most often, the standards and requirements used for human and veterinary drugs set by the Ministries of Health and Agriculture are adopted and used. Data needed includes information on efficacy against target organisms, toxicity to non-target organisms, residues in food animals, human safety, and environmental safety.

In Japan, drugs for domestic animals including those for fisheries are controlled by the "Pharmaceutical Affairs Law". Guidelines are available for the quality, safety, and efficacy criteria for the data. Data are submitted to the Central Pharmaceutical Affairs Council for review and then to the Subcommittee on Drugs for Fisheries, Subcommittee on Residues, Special Committee on Veterinary Medicinal Products, and finally to the Executive Committee. If the application is successfully reviewed by these committees, it is finally approved by the Ministry of Agriculture, Forestry and Fisheries. The approval of drugs is regulated under Good Manufacturing Practices (1991), Good Laboratory Practices (1987), Good Clinical Practices (1997), and Good Post-Market Surveillance Practices (1997). Japan estimates that its annual losses from diseases are $125 million with 70% of these losses from saltwater fish. The most common infections occur on yellowtails. Generally, it takes one year from application to approval.

In the Philippines, evaluation and control of veterinary drugs and related products are enforced by the Department of Agriculture through the Bureau of Animal Industry and the Department of Health through the Bureau of Food and Drugs. The Fertilizer and Pesticide Authority of the Department of Agriculture is responsible for registration and classification of pesticides, fertilizers, soil conditioners, and microbial inoculants. In Thailand, drugs are controlled by the Department of Fisheries. In China, imported drugs are approved for use by the Animal Drugs Examination Commission, Ministry of Agriculture.

AUSTRALIA--Applications can be submitted to the National Registration Authority to seek exemptions from the need for registration for a number of chemicals used by the aquaculture industry. A formal exemption means that the aquaculture industry can legally use a particular drug or chemical for certain specified uses without the need to get it formally registered or permitted. The following chemicals have been granted exemptions from the need for registration as a result of this process: calcium carbonate, calcium hydroxide, calcium oxide, calcium/magnesium carbonate, calcium sulfate, zeolite, aluminum sulfate, ferric chloride, inorganic fertilizers, and organic fertilizers.

A Minor Use Permit (MUP) is a short term permit given for approximately 12 months on the basis that not enough information or resources are available to sustain a full application for registration and the quantities used are relatively small. Several drugs fall in this category. After enough data are generated, the drug is approved.

EUROPE--Because the definition of a Veterinary Medicinal Product (VMP) is wide and in practice covers all the health and hygiene products that are to be used on animals, most of the very old and well known compounds that have been used over decades in aquaculture had to be reviewed, at least for the Maximum Residue Limit (MRL) assessment (Council Regulation EEC/2377/90). Old and new substances have to be defended by a sponsor who should complete the required studies, and submit a dossier to the European Commission (EC). All the existing substances which had not been given an MRL were supposed to be banned from animal usage after January 1, 1998. In fact, the EC has extended that deadline until December 1999 for those substances for which dossiers were submitted before January 1, 1996 (Council Regulation, March 1997 NE 434/97).

The result of the MRL assessment is the classification of substances into four annexes, when the MRL is fixed; when the substance does not need to be subjected to an MRL (e. g., VMPs only intended of use on pets or considered as safe in general); when a temporarily MRL is fixed; and Annex IV, when no MRL can be fixed. This latter classification results in a complete ban of the substance from the animal health usage.

The corresponding world-wide regulation relates to the Codex Alimentarius (WHO and FAO jurisdiction). Very similar files to those prepared for the European evaluation have to be submitted. Ultimately, when a food animal substance is finally approved by the Codex Alimentarius, it should provide a world-wide guarantee of its innocuity and safety.

The MRL status is also the first step to Marketing Authorizations (MAs) applicable to VMPs in each Member State. In fact, at present, no new MA application will be evaluated by the national regulatory authorities prior to the publication of the corresponding MRL. Furthermore, the question is now discussed to allow field clinical trials on a future VMP only after an MRL is published on the active ingredient. The MA authorization is given for a finished VMP, by a national regulatory authority after the review of a file that contains an additional package of information on the product, on target animal safety tests, the Environmental Risk Assessment (ERA), and the efficacy of the product (Council Directives 81/851, 81/852, 92/18). It also states the withdrawal times for the product.

Over the past years, there has been now an increasing concern on the innocuity of VMPs towards the environment. A note for guidance (EMEA/CVMP/055/96-Final) was recently issued to define the specific requirements for VMPs. This will result in a significant additional cost to the data package of all VMPs. This ERA now represents the first priority for drug registration issues in Europe, since good progress has been made on MRLs.

Fish oral drugs, external antibacterial and antiparasitic compounds, can be considered as full VMPs. In accordance with the CVMP guideline (EMEA/CVMP/153a/97), Salmonidae are being considered as a major animal food producing species which implies that a full MRL dossier should be submitted.

The MRL regulations created a big concern over the past years in the aquaculture industry, particularly when some widely used drugs were banned such as the nitrofurans and chloramphenicol and placed in Annex IV. Meetings and workshops were organized with various regulatory authorities, national and European, producers, scientists, veterinarians, and the pharmaceutical industry to discuss the future availability and non-availability of drugs for the aquaculture industry.

Today, the fish MRL situation does not look as critical as it once was, although several compounds have not received their final MRL. The CVMP has recently proposed a pragmatic approach for the establishment of fish MRLs in Europe (EMEA/CVMP/153b/97-Final) that was just approved (September 11, 1997). No additional metabolism studies should be required on fish if they have been already carried out on major food animal species. Only cold residue depletion studies would then have to completed. Furthermore, if an MRL is given on Salmonidae, it should also apply for the other fish species raised in Europe. When this proposal enters into application, some of the molecules will be given an MRL on fish very soon.

NORTH AMERICA (UNITED STATES AND CANADA)--The U.S. Food and Drug Administration's Center for Veterinary

Medicine (CVM) regulates the manufacture, distribution, and use of animal drugs. CVM is responsible for ensuring that drugs used in food-producing animals are safe and effective and that food products derived from treated animals are free from potentially harmful residues. CVM establishes tolerances and withdrawal times as needed for all drugs approved for use in food-producing animals. CVM reviews New Animal Drug Application data generated through an Investigational New Animal Drug (INAD) exemption. CVM allows other than pharmaceutical firms to be sponsors of INADs under a compassionate policy.

The U.S. drug approval process is analogous to that in Canada, requiring approval of manufacturing, human food and target species safety and efficacy sections. There are specific differences between the requirements; in particular, the United States has a specific requirement for an environmental assessment and CVM will accept product data packages in phases. The Canadian Bureau of Veterinary Drugs requires that a submission be complete before it can be reviewed and has recently introduced cost recovery fees.

The CVM understands the need to work with the aquaculture industry to resolve the drug availability situation. Over the last few years, CVM has developed several programs that clarified data requirements and created flexibility in the approval process. One of the major programs initiated by CVM was a formal low regulatory priority (LRP) status for certain drugs. Currently, 18 unapproved drugs have been designated as low regulatory priority for enforcement purposes. These drugs include acetic acid, calcium chloride, calcium oxide, carbon dioxide gas, Fuller's earth, garlic, hydrogen peroxide, ice, magnesium sulfate, onion, papain, potassium chloride, providone iodine compounds, sodium bicarbonate, sodium chloride, sodium sulfite, urea and tannic acid. The CVM considers the available information on their safety to humans and the environment, when the drugs are used under defined conditions, to be an adequate basis for deferring any regulatory action against their use. CVM views LRP drug status as a tenuous situation because they are still unapproved drugs. As a result, CVM advocates that LRP drugs go through the approval process to become legal products.

No information was presented on the drug approval and vaccine licensing programs of the various countries in South America. The only information available is that the agency in charge of approvals in Chile is the Fisheries Health Department, National Fisheries Service.

Aquaculture Vaccine Licensure in each Country or Continent

ASIA--Several vaccines are used in Asia (excluding Japan) but only three are commercially available--Vibrio parahaemolyticus vaccine (Vibrogen-S) for shrimp, Penaeid multivalent bacterin (P.M.B. vaccine) for shrimp, and Vibrio anguillarum bacterin for fish. Japan has three licensed vaccines.

AUSTRALIA--There is only one licensed vaccine in Australia and that is through the Department of Primary Industry and Fisheries in Tasmania for Atlantic salmon and rainbow trout.

EUROPE--A critical situation has been created on fish vaccine availability in Europe. New vaccines have to meet the requirements which are defined in the Council Directive 92/18 (1992). In addition, all vaccine applications which were previously accepted in each Member State have to be reviewed in the scope of this new 92/18 format. Meeting the requirement of quality standards for vaccine manufacturing (Good Manufacturing Practices) has become one major priority of vaccine manufacturers.

Fish registered vaccines are mainly antibacterial vaccines, involving the complete inactivation of the bacterins. In particular, injectable vaccines against furunculosis, were successfully developed in the 1990=s in United Kingdom (UK) and Norway. They have significantly contributed to the better results obtained in salmon farming in UK and Norway since 1992, resulting in a major drop of antimicrobial use in these countries where more than 90 % of the farmed fish are vaccinated.

Most fish vaccines are available in UK for salmon through a complete licence (MA) or an Animal Test Certificate (ATC), and meeting the major needs of the farming industry. For the other farmed fish in Europe, the availability in vaccines remains very low. These markets are much less attractive to vaccine manufacturers. New valences and new application methods are to be developed. To compensate this shortage of licensed fish vaccines in some EU countries, autogenous vaccines have been widely used.

NORTH AMERICA (UNITED STATES)--The Animal and Plant Inspection Service (APHIS) of the U.S. Department of Agriculture regulates all veterinary biologics produced in, shipped into, or exported from the United States.

Priority Aquaculture Drug Needs for Each Country and Continent

ASIA--Many drugs need to be approved in the Asian countries other than Japan. Japan needs just a few drugs to complete their medicine chest.

AUSTRALIA--Minor Use Permits have been issued for seven antibacterials, four antiparasitics, four antifungals, two anaesthetics, three spawning aids, one immunostimulant, and one piscicide (not listed in Table 3 because it is not a drug).

NORTH AMERICA (CANADA)--The status of approved therapeutants in Canada was very limited prior to the implementation of the Salmon Health project. Industry needs were assessed through a national survey that identified five national priorities. Working with manufacturers and regulators, several milestones were met in a relatively brief period of time. The list of four licensed compounds was expanded to seven with temporary registration for an additional three compounds which were under investigation. Presently data is being reviewed or developed, in co-operation with the sponsors, for new drug submissions for remaining priorities

NORTH AMERICA (UNITED STATES)--Major progress was made under various partnerships and compassionate Investigational New Animal Drug (INAD) exemptions toward gaining at least 20 new or amended New Animal Drug Applications (NADA) within at least the next three years. In addition, there are drugs that are of interest that may gain approvals beyond three years.

Researchers, Facilities, and Partnerships for Drug Approvals

ASIA--There are several groups involved in drug research and approval process for aquaculture species. In many countries in the region, the Ministry of Health and the Ministry of Agriculture is most often the approving authorities. Research institutes and universities conduct research on toxicity studies, drug screening, efficacy trials, and human safety trials. The country and the institution(s) involved are as follows:

1. Bangladesh--Ministry of Livestock and Fisheries
2. Cambodia--Department of Fisheries
3. India--Ministry of Agriculture; Indian Council of Agricultural Research, Central Institute of Freshwater Aquaculture; Central Institute of Brackish water Aquaculture; Central Marine Fisheries Research Institute; College of Fisheries, University of Agricultural Sciences; Cochin University of Science and Technology; University of Madras; and Andhra University
4. Indonesia--Drugs and Food Control, Ministry of Health; Veterinary Drugs Control Division, Directorate of Animal Health, Ministry of Agriculture; Department of Fisheries, Ministry of Agriculture; Research Institute for Freshwater Fisheries
5. Korea--Animal Health Division, Ministry of Agriculture, Forestry and Fisheries
6. Malaysia--Ministry of Health; Ministry of Agriculture; Veterinary Research Institute; Fisheries Research Institute, Department of Fisheries; University Putra Malaysia (Faculty of Applied Science & Technology, Faculty of Veterinary Medicine and Animal Science); Universiti Sains Malaysia (Faculty of Zoology)
7. People's Republic of China--Surveillance Institute of Food Hygiene, Ministry of Public Health; Bureau of Fisheries, Ministry of Agriculture; Institute of Hydrobiology, Academia Sinica
8. Philippines--Bureau of Food and Animal Drugs, Department of Health; Bureau of Animal Industry,

Department of Agriculture; Fertiliser and Pesticide Authority, Department of Agriculture; Southeast Asian Fisheries Development Center, Aquaculture Department (SEAFDEC/AQD); Bureau of Fisheries and Aquatic Resources (BFAR); University of the Philippines (College of Fisheries, Institute of Biology, Institute of Marine Science, College of Veterinary Medicine); Central Luzon State University (College of Fisheries); Mindanao State University (College of Fisheries)

9. Singapore--Ministry of Health; Primary Production Department, Ministry of National Development; Veterinary Public Health Laboratory; National University of Singapore
10. Sri Lanka--Veterinary Drug Control Authority, Ministry of Livestock and Animal Production; University of Peradeniya (Fac. of Veterinary Medicine)
11. Taiwan--Feed and Drug Division, Animal Industry Department, Council of Agriculture; Taiwan Fisheries Research Institute; National Taiwan University (Faculty of Zoology)
12. Thailand--Ministry of Public Health; Ministry of Agriculture and Cooperatives; Department of Fisheries; Aquatic Animal Health Research Institute (AAHRI); National Institute of Coastal Aquaculture (NICA); Feed Quality Control and Development Division; Kasetsart University (Faculty of Fisheries, Fac. of Veterinary Medicine); Chulalongkorn University (Fac. Veterinary Medicine, Fac. of Marine Science); Mahidol Medical University
13. Vietnam--Invention and Discovery Department, Ministry of Health; National Evaluation Council; Research Institute for Aquaculture; Nhatrang Fisheries University; Thuduc Agriculture University; Cantho Agriculture University

AUSTRALIA--Australia has established a Working Group on Aquaculture Drug Registration and has a co-ordinator to submit applications. There are no groups in Australia specifically involved in aquaculture drug approval research; however, there are many government, university and private facilities that would be capable of undertaking research relevant to this issue. One possible project is a co-ordinated effort between Australia and South East Asia for a series of residue studies on priority drugs. These studies would be in key indicator species (e.g., coldwater finfish, tropical finfish, prawn species, and temperate water marine species).

EUROPE--European funding is available through grants as (1) the SME-SMI (Craft) which associate private European SME or SMI and scientific teams. The SME-SMI are fish producers, fish vaccine or drug manufacturers, fish feed producers, and (2) the other European grants, AIR, FAR, who are mainly associating different scientific teams, but can also include a private partner.

The goals of these partnerships are not directly the development and registration of drugs and vaccines, but they certainly contribute to put the pharmaceutical firm on the right track if one firm is included. The fact that many of the studies conducted under these European grants would not meet the quality standards that are required for a registration (GLP, GCP) may limit their use in a registration file.

Another type of partnership has been seen in Europe, particularly in UK, and appeared to be very successful. These partnerships are created between the fish producer associations (SSGA and BTA), public or private institutes, drug and vaccine manufacturers. Among these partnerships, the most remarkable successes are probably the development of furunculosis vaccines, the development and registration of sea lice compounds such as dichlorvos, azamethiphos, and di-or te-flubenzuron. Present funded work relate to RFTS, sea lice vaccination, PKD control, and a malachite green replacement.

In France, such collaborations are often done under public funds, national or local, and private funds (farmers, fish feed manufacturers, and/or the pharmaceutical industry). For example, the pharmacokinetics of oxolinic acid on turbot was studied under such an arrangement. In order to get a fruitful collaboration, each of the associated partners must understand that the first aim of these funded projects is to get a product registered. This means that a fish vaccine or drug firm should be associated to the project. This partner will co-ordinate the whole project as it has a direct interest in it, carry out the registration steps onwards, and deal with regulatory authorities. To get a clear interest of the pharmaceutical industry in these partnerships, it is then crucial that it has the ownership of the results, the rights to use them for registration and commercial purposes.

Fish producers will co-operate to this type of projects if only there is a * desperate + need of a solution. Diseases due to sea lice on salmon, and PKD/RTFS on trout in UK were typically in this situation. Projects have to be clearly identified in a priority list. On an European scale, it might become difficult to get a clear picture because of the diversity of the situations. On local funded projects, it is important to define the terms of the partnership.

If French sea bass producers are ready to fund the development and registration of a product, how would they accept that these funds will actually serve a registration in Greece ? Similar concerns may be expressed between salmon producers of Scotland, and Norwegian applications.

The scientific community has a direct interest on these projects only if they can publish in agreement with the sponsors, which may not be the case. Furthermore, this work has to be carried out under GLP or GCP, which fact results in a significant study cost increase, and implies that the studies could only be carried out in fully GLP certified contract research facilities. At present, only a limited number of these facilities exist for fish work in Europe, mainly in Norway and in UK.

NORTH AMERICA (CANADA)--Salmon Health was developed to assist pharmaceutical companies to compile data submission dossiers to meet the requirements of regulatory and licensing agencies. Salmon Health is a self-funding program of the Canadian Aquaculture Industry Alliance working to ensure that aquaculturists have access to safe and effective fish health management tools in an appropriate regulatory environment. The goal of Salmon Health is to bring the needs of producers to manufacturers and regulators, bringing the latter two together in way that meets the needs of all three. Funding for the Salmon Health project comes from direct contributions from supporters who share the same objectives as Salmon Health. These supporters come from the aquaculture sector, manufacturing sectors (pharmaceutical and feed) and government agencies. Funding from all of the supporters is channelled into the following efforts:

(1) determining therapeutant priorities,

(2) identifying candidate compounds that meet these priorities,

(3) documenting therapeutant market potential,

(4) collating available information from a world wide body of scientific knowledge, and

(5) working with pharmaceutical companies to draft plans for new drug submissions.

Future issues that need to be addressed include:

(1) continuing to foster an environment which encourages manufacturers and government to develops new drug submissions,

(2) developing government and industry partnerships programs to assist with the financial cost of new drug submissions,

(3) international harmonisation of drug submission requirements for new drugs,

(4) international harmonisation of vaccine licensing and batch quality assurance requirements,

(5) international equivalency with respect to maximum residue limits,

(6) fish pharmacokinetic data with respect to new drug submissions (e.g., crop or species grouping, differences between fresh and salt water, effect of temperature, better pharmacokinetic models), and

(7) developing therapeutant use standards compliant with Hazard Analysis Critical Control Point (HACCP) protocols.

While development of safe and efficacious therapeutants remains the primary goal, increasing emphasis is placed on ensuring that the regulatory framework is in place to meet the prime directive. One issue that relates to this is the international harmonisation of drug submission requirements between countries for both therapeutants and vaccines and vaccine quality assurance requirements. In many instances these requirements are very similar, yet different enough that manufacturers have to duplicate efforts which further detracts from financial incentives to bring products to the market. Where residues are concerned, equivalence of maximum residue levels (MRL) facilitates movement of products across borders by ensuring similar therapeutant standards are developed and used. A lack of appropriate and accurate fish pharmacokinetic models exacerbates issues relating to setting minimum MRLs. Concurrent with the harmonisation of MRLs is the development of HACCP based processing and production protocols which are becoming internationally recognised food processing safety standards.

The Salmon Health model has been successful, largely due to the partnerships developed between industry, manufacturers and government agencies. In fact Salmon Health receives support from producer associations that represent 90% of North American farmed salmon production by volume, value and number of species. In the span of approximately five years the number of approved therapeutants for salmon farmers in Canada has increased from three to six, with temporary registration of three additional compounds. In the first three years of the program, many of the objectives were met in providing therapeutants to meet priorities set by the industry. As the program evolves, new priorities are being established with increasing focus on vaccine development and therapeutant regulatory and fish health management issues.

NORTH AMERICA (UNITED STATES)--Public and private aquaculture organisations are forming coalitions and appointing regional INAD co-ordinators. These new coalitions are facilitating the approval process by consolidating and disclosing compassionate INADs, developing new compassionate INADs, helping to define food/non-food fish, life stages, and crop grouping for drug-specific applications, seeking Low Regulatory Priority status for numerous drugs and chemicals, and exploring strategies for funding drug approval research.

IAFWA Project--In 1994, the National Biological Service (NBS), U.S. Fish and Wildlife Service (FWS) and International Association of Fish and Wildlife Agencies (IAFWA), on behalf of the 50 states, developed an initiative to work co-operatively to fund and carry out the research required to gain approval of eight high priority drugs (Aqui-S7 or benzocaine, chloramine-T, copper sulphate, formalin, hydrogen peroxide, oxytetracycline, potassium permanganate, and sarafloxacin) and to demonstrate the concept of crop grouping. This unprecedented partnership is one of the largest and most important agreements ever forged on behalf of fish management, production, and disease control. The IAFWA Project began July 1, 1994 and presently is envisioned to extend until June 30, 1999. At its inception, 39 states agreed to contribute funds to the initiative for five years but, to date, 37 states are participants.

UMSC. The Upper Mississippi Science Center (UMSC), la Crosse, Wisconsin is one of two federal laboratories involved in the IAFWA Project. UMSC has the combination of uniquely designed laboratory facilities for specialised fishery drug and chemical research, a balanced blend of highly trained professional staff, state-of-the-art instrumentation, laboratory capabilities to test a diverse range of aquatic organisms, an established and recognised quality assurance program, and a proven record to develop approval and registration information for aquaculture drugs and fishery pesticides.

SNARC. The Stuttgart National Aquaculture Research Center (SNARC), Stuttgart, Arkansas, is the other federal laboratory involved in the IAFWA Project. SNARC has begun a recent program of working on drug approvals for the commercial aquaculture industry. The laboratory is hiring additional scientific staff, and recently built a new laboratory building that will emphasise drug approval research.

BPA/FWS Erythromycin Registration Project. In 1989, the Bonneville Power Administration (BPA) provided funding for the first comprehensive, publicly funded drug registration project to conduct the necessary research and assemble drug registration packages for use of erythromycin as a feed additive and injectable treatment in salmonid fish culture. In 1992, at the request of the Pacific Northwest Fish Health Protection Committee, this drug registration effort was expanded so that field trials with erythromycin could be conducted at salmonid culture facilities throughout the Pacific Northwest. The University of Idaho conducted all of the laboratory studies needed for approval and has searched for a pharmaceutical sponsor.

Universities. Several universities has been involved in the drug approval process in recent years in the United States. Most of their efforts have centred on efficacy and target animal safety studies. Some of them include Auburn University, Auburn, Alabama; University of Arizona, Tucson, Arizona; University of Florida, Gainesville, Florida; University of Idaho, Moscow, Idaho; Southern Illinois University, Carbondale, Illinois; Louisiana State University, Baton Rouge, Louisiana; Mississippi State University, Mississippi State, Mississippi; Rutgers University, New Brunswick, New Jersey; Cornell University, Ithaca, New York; Ohio State University, Columbus, Ohio; Texas A & M, College Station, Texas; and University of Wisconsin-Madison, Madison, Wisconsin.

U.S Fish and Wildlife Service--The U.S. Fish and Wildlife Service (FWS) established a national Investigational New Animal Drug (INAD) office in Bozeman, MT in 1994 that is co-ordinating all FWS-INAD related activities (e.g., protocol development, data collection and analysis). These activities involve 12 individual INAD exemptions, 90 facilities, and approximately 300 INAD units.

Western Regional INAD Project--In November 1993, the Western Regional INAD Project was established as a government-private sector partnership to sponsor and manage clinical field trials for three therapeutants (oxytetracycline, chloramine-T, and formalin) at 218 facilities operated by six state conservation agencies, three tribal groups, and 24 private fish producers in Alaska, California, Idaho, Montana, Oregon and Washington. The project terminates on December 31, 1997 and there is hope that these INADs will be extended by other sponsors.

States--Twenty states have 61 individual INAD exemptions for eight therapeutants, one anaesthetic, and three spawning aids. Many states have offered to consolidate their INADs with other states, and in some cases, with the private sector.

Veterinarians--Veterinarians have increased their involvement in aquaculture and have teamed up with fish health professionals working for state and Federal agencies. This process has been aided by an Aquaculture and Seafood Advisory Committee established by the American Veterinary Medical Association (AVMA) and by discussions with the American Fisheries Society's Fish Health Section.

JSA. The Joint Subcommittee on Aquaculture (JSA) recognised that additional approved safe and effective aquaculture drugs are necessary to manage production fish health and responded by creating the Working Group on Quality Assurance in Aquaculture Production (Working Group) in 1990. The mission of the Working Group is to increase understanding of, and compliance with, Federal requirements regarding drugs, chemicals, and biologics used in aquaculture production in the United States. The Working Group advocated the establishment of a National Coordinator of Aquaculture New Animal Drug Applications in 1995. The Working Group met in September 1996 and developed a strategic plan for the next five years. Two of the goals and objectives in the five-year plan were as follows: (1) to encourage pharmaceutical, chemical, and vaccine industry involvement; and (2) promote the participation of domestic and international agencies and producer groups in the activities of the Working Group on Quality Assurance in Aquaculture Production.

CVM. The Center for Veterinary Medicine (CVM) understands the need to work with the aquaculture industry to resolve the drug availability situation. Over the last few years, CVM has developed several programs that clarified data requirements and created flexibility in the approval process. The major programs are formal compassionate investigational new animal drug (INAD) exemptions, formal low regulatory status for certain drugs, flexible labelling, innovative approaches to efficacy and target animal safety studies, encouragement of crop grouping research, and encouragement of disclosure and consolidation of INADs. The Joint Canadian-United States Workshop on Jurisdiction of Sea Lice Treatment and Control established a joint Canada and United States Aquaculture Working Group to institute international harmonisation activities. This means that data could be shared and certain requirements for all drugs could be harmonised so that there could be joint submissions leading to approvals being granted simultaneously in both countries.

The International Harmonization Workshop for Aquaculture Drugs/Biologics, held in Seattle, Washington on February 24, 1997, created an educational forum to exchange information and identified issues between public and private sectors and international organisations with the goal of initiating follow-up strategies to advance harmonisation of drug maximum residue levels, aquaculture drug approval standards, and biological licensure. Two major animal drug laws and follow up regulations were recently implemented in the United States that have an impact on the approval of aquaculture drugs: The Animal Drug Availability Act of 1996 and the recently finalised regulations for the Animal Medicinal Drug Use Clarification Act of 1994. The former codifies the efforts by CVM to facilitate minor use drug approvals and the latter clarifies extra-label use in aquaculture.

ROUND TABLE DISCUSSION

The speakers and the audience participated in a round table discussion in Edinburgh, Scotland, on the aquaculture drug and vaccine issues that need addressing from a world-wide and country perspective. The first topic was a review of what had been presented.

It was agreed that the approved and priority drug and vaccine needs had been identified but that not much had been presented regarding market surveys. Some research facilities and partnerships for funding were identified; however, no protocols were discussed.

A major area of discussion at the Round table centred on identifying standards, regulations, and guidelines that needed consideration as they related to drug approvals and vaccine licensure. The topics included the need to (1) reduce costs of fees for European Union products, (2) clarify the minor species cascade effect, (3) standardise and extrapolate MRLs, (4) advocate for the use of inherent withdrawal times and early life stage models, (5) address the differences in enforcement policies between countries, (6) create economic incentives for pharmaceutical firms to enter the aquaculture arena, and (7) develop biodiversity mechanisms.

Another area of discussion centred on the development of advocacy groups, communication mechanisms, and educational programs. The topics included the need to (1) encourage producers to buy only licensed products, (2) encourage groups to develop and use quality assurance programs, (3) develop integrated fish health management programs, (4) work with feed manufacturers on aquaculture feed issues, and (5) consider post-approval monitoring for environmental effects, presence of residues, and disease resistance.

The group agreed that communication mechanisms need to be in place to keep all players informed on drug approval and vaccine license issues. They agreed that the information presented at this workshop needed to be available in a publication and presented at the 1998 International Harmonisation Workshop on Aquaculture Drugs and Biologics to be held in Las Vegas, Nevada (USA) in February 1998. Mail groups need to be set up via e-mail to provide for a forum for information exchange and advocacy. Web sites need to be set up to help in the transfer of information.

CONCLUSIONS

A start have been made toward developing an international effort toward aquaculture drug approvals and vaccine licensure and toward establishing harmonised guidelines and standards for facilitating registrations world-wide. Hopefully, this international effort to facilitate aquaculture drug approvals and vaccine licensures will continue to move forward through the various communication networks and committees that have been established by this Workshop and the Workshops on International Harmonisation for Aquaculture Drugs and Biologics

STREPTOCOCCAL INFECTIONS OF FISH

Streptococcus is a genus of bacteria that includes some species that cause serious diseases in a number of different hosts. A major identifying feature of Streptococcus is that they are Gram-positive (they appear purple/blue when stained using a procedure called a Gram stain). By contrast, most of the common disease-causing bacteria of fish are Gram-negative (appear pink with a Gram stain). This is important when considering treatment options, and will be discussed in greater detail below.

Streptococcal (Strep) diseases of fish are not common; however, when they do occur, significant mortality can result. Some aquatic Strep species may cause disease in humans in unusual circumstances. However, these latter species do not usually affect healthy people. In addition to bacteria in the genus Streptococcus, there are several other closely related groups of bacteria that can cause similar disease, including Lactococcus, Enterococcus, and Vagococcus. For purposes of this fact sheet, all of these bacteria and the disease itself will be referred to as Strep.

Streptococcal disease in fish was first reported in 1957, affecting cultured rainbow trout in Japan (Hoshina et al. 1958). Since then, numerous other species of fish have been found susceptible to infection, including salmon, mullet, golden shiner, pinfish, eel, sea trout, tilapia, sturgeon, and striped bass (Inglis et al. 1993). Strep has also been isolated from a variety of ornamental fish, including rainbow sharks, red-tailed black sharks (Russo, Mitchell, Yanong 2006), rosey barbs, danios, some cichlids including Venustus (Nimbochromis ("Haplochromis") venustus) and Pelvicachromis sp., and several species of tetras.

Strep infections in fish can cause high mortality rates (> 50%) over a period of 3 to 7 days. Some outbreaks, however, are more chronic in nature and mortalities may extend over a period of several weeks, with only a few fish dying each day. A typical history suggesting that Strep may be the cause of disease in a group of fish might include reports of abnormal swimming behavior, often described as spiraling or spinning. Anytime fish are observed behaving in an unusual manner, Strep should be considered as one of the possible causes. However, not all infected fish show abnormal behavior.

Keep in mind that the keys to disease prevention in fish include: a) maintaining good water quality; b) providing proper nutrition; c) keeping the environment clean; and d) quarantining new fish before adding them to an existing collection. Knowing which species are susceptible to Strep and seeking assistance for rapid diagnosis and proper therapy if a disease outbreak should occur are important ways to decrease losses.

FACTORS CONTRIBUTING TO DISEASE

Most infectious diseases of fish are opportunistic. This means that the simple presence of the pathogen in the environment of the fish is inadequate to cause a disease outbreak. Other factors usually come into play such that either the pathogen has an advantage over the host or the immune system of the host is compromised in some way, increasing its susceptibility to the pathogen.

This phenomenon is often precipitated by "stress," and it is discussed in greater detail in IFAS Circular 919, Stress-Its Role in Fish Diseases. Stress often plays a significant role in outbreaks of infectious disease in fish populations. Some stressors that have been associated with Strep outbreaks include high water temperatures (e.g., during the summer), high stocking densities, harvesting or handling, and poor water quality, such as high ammonia or nitrite concentrations. Bacteria that cause disease in fish are often categorized based upon their perceived or documented virulence (ability to cause disease); however, these are inexact characterizations.

In reality, individual bacterial strains often differ in their virulence, and many bacteria fall somewhere within a wide spectrum of virulence, with true opportunists being the least virulent and primary pathogens being considered the most virulent. For example, the most common bacterial pathogen in freshwater fish, Aeromonas hydrophila, is considered a truly opportunistic pathogen, because it is relatively common in the aquaculture environment (hence, the term "environmental bacteria") and typically does not cause disease in healthy, well-maintained fish populations.

Strep, on the other hand, does not seem to be a truly opportunistic pathogen, as it can be more aggressive than many other environmental bacteria.

In one experimental study (Ferguson et al. 1994), populations of zebra danios and white cloud mountain minnows exposed to high concentrations of Strep in the water experienced 100% mortality within 2-4 days of exposure. Consequently, it is important that infections be quickly identified and managed to prevent major losses.

CLINICAL (OBSERVABLE) SIGNS

Affected fish may exhibit one or more of the following clinical signs, depending upon the species: erratic swimming (such as spiraling or spinning); loss of buoyancy control; lethargy; darkening; uni-or bilateral exophthalmia ("pop-eye" in one or both eyes); corneal opacity (whitish eyes); hemorrhages in or around the eye, the gill plate, base of the fins, vent/anus, or elsewhere on the body; ascites (i.e., distended abdomen/dropsy/bloating); and ulcerations.

In some cases, the fish may show no obvious signs before death. Of the signs listed above, hemorrhage, pop-eye, spinning, and rapidly progressing mortalities are among the most frequent observations.

NECROPSY (INTERNAL) FINDINGS

Internal examination may reveal the presence of blood-tinged fluid in the body cavity, an enlarged reddened spleen, pale liver, as well as inflammation around the heart and kidney. Many Streps infect the brain and nervous system of fish, explaining the erratic swimming frequently observed in infected fish.

DIAGNOSIS AND TREATMENT

Most common opportunistic fish bacteria, when viewed with a microscope at high power, appear pink after a procedure known as Gram staining. These pink-staining bacteria as a group are called Gram-negative. Most of these are also rod-shaped. By contrast, Streps appear purple/blue after Gram staining; purple/blue-staining bacteria as a group are considered Gram-positive.

Streps are also typically more spherical or oval in shape (coccus meaning ball-or sphere-shaped). A tentative diagnosis of Strep can be made from the history and clinical signs, necropsy findings, and identification of Gram-positive bacteria from stains of impressions (produced by blotting sections of fresh tissues onto a glass slide) from the brain, spleen, kidney, or liver. Strep should be highly suspected if fish exhibit abnormal swimming behavior, pop-eye, hemorrhages, and rapid and severe mortalities, and Gram-positive cocci are found in brain, kidney, and/or other organs.

A confirmed diagnosis requires culture of internal organs, especially the brain and kidney, followed by identification of the bacterium. Blood agar is the medium of choice for growing Strep. Identification is important for determining treatment options. There are a variety of commonly used antibiotics that generally work best against Gram-negative bacteria, and others that work best against Gram-positive bacteria, including Strep.

Ideally, after the bacterium has been identified from a sick fish, a sensitivity test should be conducted to select the most effective antibiotic to use. Typically, Gram-positive bacteria, including Strep, are susceptible to erythromycin. One oral dosage rate that has proven effective in the field is 1.5 grams of erythromycin per pound of food, fed for 10 to 14 days. Amoxicillin has also been shown to be effective against Strep in experimentally infected tilapia and sunshine bass at an oral dosage rate of 80 mg/kg body weight (= 3.6 grams per pound of food) fed for 8 to 12 days (Darwish and Ismaiel 2003; Darwish and Hobbs 2005).

However, sensitivity testing will confirm whether or not one of these is the antibiotic of choice when an outbreak is in progress. Consequently, the assistance of a fish health specialist or diagnostic laboratory is highly recommended. If you raise food or game fish, there are restrictions on which antibiotics can be used legally to treat these animals. Consult with a fish health specialist before attempting to treat fish with a potentially illegal drug that could result in tissue residues. A fish health specialist can also recommend the proper dosage regime for a particular antibiotic and situation.

PREVENTION

Prevention of disease is always preferable and more profitable than treatment of disease outbreaks. Preventive medicine programs should be designed to minimize stress (see IFAS Extension Circular 919, Stress-Its Role in Fish Diseases); maintain the best water quality possible (see IFAS Extension Circular 715, Management of Water Quality for Fish); and minimize exposure to infectious agents by following appropriate disinfection and sanitation protocols (see IFAS Extension fact sheet VM-87, Sanitation Practices for Aquaculture Facilities).

Although Strep does seem to occur more frequently at warmer temperatures, it can occur at any time of the year.

Studies in marine systems in Japan (Kitao et al. 1979) indicate that Strep may be present in salt water and mud, with higher incidences in the water during summer months. This implies that, in aquaculture systems some Streps that occur naturally in the environment may become endemic (established within the farm) and cause disease on a periodic basis. As a result, Strep infections may be a cyclic, recurring problem especially during periods of high stress.

If you suspect that you have a group of fish with a Strep infection, the affected population should be isolated from all others. If possible, have dedicated equipment (nets, siphon hoses) for these fish, and be sure to follow each use with an antibacterial dip such as benzalkonium chloride or Virkon® (follow manufacturer's instructions). Strep infection can spread rapidly in a group of fish exposed to water contaminated with the bacteria. In addition, Strep can spread in a group of fish through oral routes, including cannibalism of infected fish, or through feeding of infected fish food. Remove dead fish as soon as possible to minimize oral transmission. Live or unprocessed (fresh or frozen) foods should be considered a possible source of exposure, and they should be submitted to a microbiology lab for bacterial culture when an outbreak is being investigated.

Immunostimulants added to the feed, such as beta-glucans and nucleotides, have been shown to increase survival for infected redtail black shark (ornamental fish) populations (Russo, Yancng, Mitchell 2006).

Vaccines may be useful for facilities that have continual or cyclic outbreaks of Streptococcosis. Autogenous vaccines (vaccines developed for a specific facility, targeting a specific bacteria isolated from a disease outbreak at that facility) have been shown to be effective under certain conditions. Commercial vaccines may also be available for use within the next few years.

Species of Streptococcus that May Cause Disease in Fish

There are many different species of Strep that cause disease in fish. Many of these have not been fully characterized. Identification features common to Strep species that cause disease in mammals are frequently not applicable to identification of those species that cause disease in fish. Many of the Streptococcus species that are pathogenic (disease causing) to fish are found naturally in the environment and may become endemic to a farm. There is not enough information available for most to determine which species or strains of Strep are more pathogenic to fish.

Examples of Strep species that have been associated with disease in fish include Streptococcus difficilis, isolated from fish in Israel; S. milleri, isolated from the kidney samples of koi observed with external ulcerations; and S. parauberis, isolated from farmed turbot in northern Spain (Austin and Austin 1999).

Streptococcus iniae is another species that has been isolated from diseased fish and mammals (Austin and Austin 1999). Other Gram-positive bacteria that are closely related to Strep and which have been associated with disease in fish include,

Lactococcus garvieae, isolated from diseased eels and yellowtail in Japan (Austin and Austin 1999); L. piscium; and Vagococcus salmoninarum, isolated from diseased rainbow trout.

DISEASE IN HUMANS

Streptococcus iniae is known to cause disease in both fish and mammals, including humans (Austin and Austin 1999). Streptococcus iniae was first isolated from an Amazon River dolphin in the 1970s. In 1994, S. iniae was reported as the disease agent from an outbreak in tilapia hybrids. However, it was later determined that this Strep was identical to S. shiloi, first isolated from tilapia from Israel in 1984. In 1991, the first recognized case of S. iniae in humans occurred in Texas (MMWR 1996). In 1994, a second case occurred in Ottawa, Canada. However, the source of infection for both cases was not determined (MMWR 1996). In 1995, a group of Canadian doctors described Streptococcal disease in a group of older fish handlers, of whom at least four individuals had additional underlying ailments, including diabetes and/or rheumatic heart disease (Weinstein et al. 1997). They were known to have injured their hands while preparing whole raw fish, primarily tilapia, and subsequently became ill. The most common type of infection (8 of 11 patients) was cellulitis (infection of the connective tissue) of the hand, causing fever in these patients. One patient, who had other health risks (heart disease, kidney disease, diabetes, and arthritis), developed a more serious infection. All cases were treated with antibiotics and resolved.

Although humans did become ill from a Strep infection originating from the handling of diseased tilapia, it is important to note that the people in this group were older (median age 69 years). Several were in debilitated health, and thus, as a group, had compromised immune systems. Also, the majority of affected individuals became ill only following puncture wounds and/or cuts to their hands.

These and other data and observations suggest that the risk of healthy humans getting Strep from diseased fish is apparently very low. Fish health experts working with tilapia producers in several states have not seen any problems resulting from farm workers handling diseased tilapia. People who are immunocompromised are at a greater risk especially if they suffer cuts or puncture wounds, and thus these individual should follow appropriate precautions, such as wearing gloves and using antibacterial soap. Although it is important to be aware of the potential risk posed by S. iniae to immunocompromised workers, information to date suggests that the risk is minimal to healthy individuals (Shoemaker and Klesius 1997).

SUMMARY

Although not as common as other bacterial diseases, Strep infections do occur in fish, and can result in significant financial loss to the fish farmer. Streps are Gram-positive bacteria. Fish populations with Strep infections typically experience moderate to heavy losses.

Because Strep frequently infects the brain, abnormal behavior, such as spinning, is often noted. Other signs include lethargy, darkening, exophthalmia ("pop-eye"), hemorrhages, ascites (dropsy/bloating), and ulcerations, although these are not necessarily specific to Strep. However, in many cases, there may be no obvious signs other than rapid mortality. Other potential causes of disease should be ruled out as part of the diagnostic process. If you suspect you may have a Strep infection in your fish, consult a fish health specialist who can isolate the bacteria from the fish, properly identify the disease-causing organism, and recommend appropriate antibiotic therapy. Vaccines and immunostimulants may become more important preventive-medicine tools against Streptococcosis in the future.

Streptococcus iniae, known to infect certain fish species, has caused disease in humans, including one group of people in placecountry-regionCanada who handled live tilapia. However, these people were considered much more susceptible to disease than the general population because of their advanced age and underlying health problems. Additionally, puncture wounds or cuts were required to initiate infection in this group. Normal, healthy persons are at minimal risk of acquiring this disease from sick fish.

11

High Value Aquaculture Production

Though the world, farmers, traders, planners and investors are watching the progress of modern aquaculture, the dynamic markets to be secured, the use of sophisticated technology, and the dramatic increases in production. It is clear that there is a successful industry, the 'star' sectors, such as shrimp and salmon, have shown strong, expanding markets,good profitability, and growth in production often more than 30% annually, which has been the envy of many other industries.

This success has been associated with increasingly sophisticated technology, and in the case of salmon, perhaps the most dramatic case, increasing mechanisation, and the emergence of 'industrial' approach to production. It has also been very closely dependant on effective market development.

High value methods of aquaculture are quite clearly defined in a range of cases, and details can be found in many of the standard texts.

SPECIES CHOICE AND PRODUCTION CHARACTERISTICS

There are many forms of aquaculture, involving a range of species, habitats, production methods and resource requirements. To assess the nature of aquaculture and its relationships with the aquatic environment, and before considering projects in detail, it is useful to consider adopting a 'systems' approach. A wide range of species is involved, though an even range is potentially cultivable.

Continuing research brings increasing numbers of species within these bounds, but only certain species are likely to be commercially viable for use. Although species may differ widely in habitat, behaviour and physiological response, there are in fact many common elements in the relationships between species, system and environment. Thus we will see that common links occur in areas such as stocking density, feeding quality, density and opportunity, water requirements, and waste production. The implication of this is that we can find many similarities across species when used in aquaculture production. In most geographical or market areas, there are usually certain clearly defined species of interest for aquaculture. The main factors to consider are usually:

Market value; high market value makes possible relatively sophisticated aquaculture techniques, and may support some research and development in early production stages. Low market value makes it difficult to consider any but the most basic methods of aquaculture.

Overall ease of culture; tolerance of crowding, hence lower capital costs per unit of capacity, and ease of feed distribution, management methods, etc. Low disease risk improves overall project risk. Good growth rate increases stock turnover, hence lowers capital cost per annual production. Good environmental tolerance reduces water requirements, and/or need for close environmental monitoring or control.

Simple feeding requirements reduces need for sophisticated, expensive feeds and feeding systems, simplifies overall management. Permits routine and large-scale production. In many cases, feeds determined system employed-eg natural feeding or artificial feeds.

Simple hatchery techniques with full life-cycle control; routine supply of fry, with reliable production planning hence lower operational costs, less risk. Year-round larval supply improves utilisation of holding facilities, particularly of hatchery. Difficult hatchery techniques require sophisticated and expensive systems.

Simple handling/harvesting requirements: permits ready examination, grading treatment, regular and reliable harvesting, development of larger-scale production systems. To illustrate the use of these criteria, Table 1 summarises the use of the most important preliminary screening criteria associated with typical higher-value freshwater aquaculture species. Certain of these, such as rainbow trout, salmon and eels relatively well-established in aquaculture, and can be adapted readily to local conditions. Others may not be fully known led at a commercial scale in the region concerned, and remain to be evaluated.

PRODUCTION SYSTEMS

Trout, eel,carp, catfish and tilapia are all being grown in various parts of the world in hyper-intensive systems, sometimes, incorporating complete environmental control, and water recycling to conserve water and allow units to be set up almost anywhere. Such systems, and other such as fish/ hydroponic units, are increasingly being proposed or marketed as complete turnkey 'food production package'. The most common units for higher value fresh water species are intensive tanks, raceways or cages.

Tanks and raceways

Various designs are used. Because of associated high capital operational costs, they are principally used in intensive, commercial-scale operations where high market prices can be realised, and in hatcheries, with high value/biomass and greater need for high-quality controllable husbandry conditions.

Tanks can be virtually any shape. Although round may be more expensive, they are superior in terms of flow characteristics and thus waste removal and water quality. Tank size is determined by economic and management considerations. Generally, larger tank is are less expensive on a per unit volume basis. Tanks much greater than 10m in diameter are difficult to design, install and manage. As a general rules, a diameter depth ratio of 5-10: 1 is desirable to ensure good cleaning.

Raceways are large, elongated tanks. Because of their shape, water flow characteristics are more even and predictable, though as water quality changes down their lengh, husbandry characteristics may be inferior. Length: width: depth ratio should be around 30:3:1. An advantage of raceways is that they can be terraced, eg down a slightly sloping site, with intermediate cascade aeration between each stage.

Water supply should be around 0.5-11 kg fish-1 min-1. Current velocities would normally be between 10 and 25 cm s-1, though this depends on species requirements, stock size, etc. Tanks normally have a peripheral, tangential inflow and a central drain, with a circular vortex water movement. This allows velocities within the tank to be controlled independently of water flow.

Raceways normally have an inflow at one end and an outflow at the opposite end. Design is critical ideally, water should have a laminar flow and approach "plug" flow conditions in which the entire water column moves at the same velocity, thus eliminating dead spots and short-circuiting and maintaining a high cleaning efficiency. The inlet should be full width and baffles may be necessary to reduce turbulence. Unlike tanks, water velocities are entirely dependent on flowrate.

Tanks and raceways can be fabricated from a variety of materials. Costs, durability and a smooth interior surface to reduce risks of abrasion damage are important. Simple, inexpensive tanks can be fabricated from sections of large dia meter concrete pipe embedded in concrete and finished inside with a skim of fibreglass or plaster. France and liner tanks and raceways are also very cheap. Otherwise GRP (fibreglass)-expensive, but light and very smooth, and brick, concrete block or reinforced concrete-cheap, but much rougher-need good plaster finishing-are commonly used. Foundations often need care, particularly for heavier units, soils need to be checked for strength, etc, and some bedding may be needed. Water supply demands are usually greater than for pond systems, either:

- higher quality, eg for hatchery supplies:
- larger quantities, continuous supply, for intensive ongrowing.

For hatchery water supplies, it is common to use a sand filter, often eg a sub-sand beach filter, a settling/reservoir tank and a series of mechanical filters, such as swimming-pool filters and cartridge filters, to remove pathogens and finer solids. This is particularly important for live feeds and early lifecycle stages. Ongrowing systems typically have a large pumping station, a distribution chamber, and pipe or channel supplies to the tank or raceway units. Pumps usually run continuously and have to be carefully designed to minimise operational costs. Distribution systems have to be eleanable, particularly wastes. Double supply lines are something used, to allow one to be cleaned white the other is operational.

Cage Systems

Cages are perhaps the most versatile and cost-effective units used for aquaculture production in estuaries, coastal regions and the open sea, they provide easily manageable holding volume at relatively low cost. The variety of cage designs has arisen out of attempts to deal with a number of (at times conflicting) design objectives.

- providing a reasonably stable shape, to avoid stressing the stock, and to provide a stable working environment:
- providing adequate water exchange to satisfy metabolic requirements of stock and remove wastes from the cage area:
- absorbing or deflecting environmental forces, to maintain the structural soundness of the system;
- providing an efficient working environment, for routine husbandry, and where equipment and materials (harvested fish, feed, tanks, and bins, etc) can be handled.
- maintaining position, to provide a secure location, free from navigation hazards, etc.

They may have specific advantages. However they cannot be developed in every location.

Comparative advantages of cage systems

Relatively low cost, particularly for larger volumes; though modern systems for exposed locations may equal tanks or raceways,

Relatively simple and fast to assemble, though larger more complex systems may require boats and lifting gear; sites may be developed quite rapidly;

Not too dependent on land availablity-only for services base, etc; but needs good quality aquatic site, with good reliable acess;

Easy to move and relocate if needed; whole assemblies complete with stock can be towed to other locations; equipment can also be sold or re-used in other sites;

Do not require water supply installation; though sites must ensure good water exchange through cage nets, avoiding local accumulation of wastes;

Relatively easy to services; nets can be lifted to gather or harvest stock; but nets regular cleaning, and disease treatment can be difficult; cage systems are also often sensitive to weather conditions;

A cage system behaves 'dynamically'-forces affecting the cage frame are transferred to the net, to other cage frames and to the mooring system, and the corresponding motion is also transferred.

This will affect both the durability of the system, and the acceptability to the stock. Cage elements tend to be designed either as flexible or as rigid stuctures; cage systems are usually a combination of these. Some basic concepts environment and stock holding are:

- theoretical carrying capacity of a cage depends, on the open net area across the current flow (where present);
- carrying capacity depends on mesh area, degree of fouling, transmission factor of netting;
- in cage groups, cages downstream of currents have poorer water flow and hence carrying capacity;
- if tidal flow is zero (eg at top or bottom of the tidal cycle), there is to water ex change;

While these are useful in estimating water exchange, there are several other factors:

- the stock itself creates water flow, usually in a vortex pattern, with new water entering the base of the cage.
- there may be convection and density currents, also local currents moving water through the cage base.

- there is some movement of water due to wave and wind action, particularly n the top 1-2m of the cage.

A further environmental factor in cage systems is the removal of wastes. Generally, ammonia and carbon dioxide can be dispersed in normal water flows. However, solid wastes (food and faeces) tend to deposit below or around the cages themselves. Their dispersion, and their probable effect on the cages will be determined by:

- their sedimentation rate;
- currents in the water column;
- local bed profile and localised currents;
- the leaching rate from the sediment;

Cage frames: are positioned on or near the surface, normally support the net, hold the flotation elements, connect to the mooring system or other cages, and may carry decking on which various loads can be carried. Some designs also incorporate a handrail. The main types are as follows:

Wool/bamboo: this is probably the simplest, cheapest and most widely used frame type in tropical areas where sites are relatively sheltered. Frames are nailed, pinned or tied together, and are usually slightly flexible. As the frames are usually quite light, they do not always incorporate a working deck.

Wood and steel: these are perhaps the most widely used types, and if properly designed, can cope with moderately severe exposure. Frames are usually square, of 7 to 15m side, and normally incorporate decking and handrail.

Steel/aluminium frames are considerably more expensive than the previous types, and are commonly used for larger cages, typically 10 to 20m. square. For exposed conditions, these systems require careful design and high-quality fabrication.

Flexible tube, usually butyl rubber, is commonly used for moderate to high exposure designs, typically in 15-20m diameter near-circular structures of 6 to 12 sides, or smaller 5-12m square frames. This method is similar in cost to steel/aluminium. They system lies in the wave zone but is so flexible that absorbed stress is minimised. The whole system is held open in shape by its mooring connections. Plastic pipe-normally PVC or ABS is used in a similar way to flexible tube, though for smaller cages-2-5m square of 5-15m diameter, which are more rigid design. These designs are approximately similar in cost with wood/steel systems. The cage system rides in the wave zone but is reasonably flexible, and allows larger waves to wash over through. Fatigue cracking and brittleness from UV exposure are among the main problems.

Where flotation is not already incorporated in the frame structure (eg plastic or flexible tube systems), it must be provided and attached to the frame.

Plastic flouts-buoys, etc can be quite useful for smaller flotation requirements; these can be lashed inside a support framework, or tied together through handles, eyes, etc for heavier loads. Drums-these are typically plastic chemical drums or steel oil drums. These are widely laid horizontal, tied between or below cage framework, as main flotation units. Depending on the amount of protection against UV or corrosion, these can be reasonably durable, though sometimes relatively expensive.

Polystyrene block-perhaps the most common flotation material, used either uncovered; or inside GRP or plastic casings, block can be sized for specific designs, and can be easily positioned in cage framework.

The main stock nets are normally suspended from within the cage frame. Typical dimensions range from 3 to 20m across, 3 to 10m deep. Most commercial operation use nets of 250 to 100m3 in volume, corresponding to about 200 to 500m^2 of netting. This is usually assembled in several panels, with suitable mounting rope, loops, and other fixings or attachments. At least 50cm of netting, sometimes of lighter grade, is extended above sea level, usually fixed to the handrail or a similar support, as a 'jumper' to prevent the escape of fish. Additional, lighter nets are commonly placed above and around the cage to deter predators.

As the nets must be handled regularly, and in most cases frequently changed, they should be relatively light and manageable, yet durable. They should not cause damage tot he stock, whether through trapping or abrasion. Mooring systems typically comprises:

- fixing devices; typically anchors, blocks or holdfasts such as shore fixing pins.
- connecting lines; usually a combination of chain, cable and rope, sometimes with additional
- components buoys, weights, etc to modifiy the elasticity of the line.
- surface gera; lines, platforms and/or buoys to which individual cages or groups of cages can be attached, and which can be used as part of the operating system for the cages.
- connecting gear; chains, shackles, pins, hinges and links, together with 'shock absorbers' such as rubber tires, buoys, etc to limit wear between specific components.

WATER DEMAND IN INTENSIVE AQUACULTURE SYSTEMS

Quantities per tonne of production vary widely, determined primarily through managed water exchange associated with intensity of production (i.e stocking density and use of feeds and fertilisers), but also through physical factors such as seepage and evaporation. It must of course be clarified that these figures do not represent actual consumption, but represent the quantities of water which have to be made available to bring about specified forms of production. The main implications are then the opportunity costs and the quality changes involved.

TECHNOLOGY OPTIONS FOR INTENSIVE, HIGH-VALUE AQUACULTURE

Yields from high-technology systems can be impressive. Can then the salmon farms of Western Europe, the recycle eel and catfish systems-indeed can be whole high-technology approach be an answer to our needs for aquaculture production? The answer, of course, is maybe, and sometimes but certainly not always'. Wet then are the critical factors-how can we decide whether to use high technology, and if so, how much and what kind to use? How should we choose the technology once we decide to use it? First some general points-maybe self-evident, but often overlooked:

- Aquaculture still relies fundamentally on principles of sound animal husbandry-good stock, the righ environment, good feeding, and management. Technology can support these but not substitute for them. Can you provide the husbandry?
- Technology can be expensive; can the market price and the projected revenues support the price of technology?
- Intensive, high technology systems in particular, require ample resources; good quality feeds, energy, good or controlled quality water, effective disease control, skilled staff and good management. Are these available?
- Off-the-shelf technology often looks (and sometimes works) better on-the-shelf. Does it really match what you need?
- Technology does not necessarily transfer well from species to species, or from location to location. Are you venturing into the unknown?
- Technology can offer better control, but at the expense of greater dependence. What happens if it goes wrong?

It always pays to ask these basic questions first. To consider some of the specific areas where technology can be used:

Systems These are several different types of high technology aquaculture system on the market today, ranging from packaged cage units, cage barges and converted ships, to package recycle systems and modular hatcheries. Generally, there are three approaches in current use.

1. Hardware-this has the advantage of being simple, relatively flexible, and if the firm is reliable, well-proven. However, the onus is usually on the buyer to decide whether the system will work in their ow ı conditions, and whether they have the resources to operate them. Reliable firms will offer advice, and should be able to offer case studies, and ideally contacts concerning similar projects. Reputable suppliers will also advice the client if their own systems are unsuitable, or whether the

client needs to change projects ideas. In some cases, the supplier may be able to work with the client to develop the appropriate system, but this depends much on the skills, resources, and attitude of the supplier.

2. Management-inclusive packages-this approach offers the advantage of management tailored to the systems, and in some cases, some form of production target or even guarantee. The supplier will usually therefore have to take greater care to ensure the system fits the local requirements. However, this approach can be relatively expensive, and may involve the client in longer-term contractual commitments which may work to the client's disadvantage. The buyer should be careful to check the competence of the actual management team proposed, and their ability to work in the local conditions. It is also important to consider how dependent the system is on the continued use of outside assistance-can local staff and resources be used? What about training?
3. Financed packages-there are two main forms; the first, where the supplier can offer financing, usually through a commercial lender or through export credit for some or all of the package; or the second where the supplier establishes a joint-venture operation with the client, sharing the risks and returns of the enterprise. In the first approach, there may be a requirement that the supplier manage the project or operate a service contract, as part of the security of the project. The approach is then generally similar to that of the previous case.

The second approach should offer the advantage of ensuring that the supplier is also interested in efficient and profitable operation this route is sometimes favoured by specialist groups with expertise and access to capital, but lacking access to sites or other critical resources. However, the use of an expensive service contract can sometimes reduce or even eliminate the specialist partner's risk, and hence their motivation for success. Obviously, the client will have to be even more careful than in other cases about determining the financial and managerial strength of their partner.

Perhaps one most difficult tasks facing a new entrant to the aquaculture industry is that of finding out exactly what typical operating or performance standards might be expected of any proposed system, and what the risks might be in choosing a particular package. For larger projects particularly, it is well worthwhile employing an independent professional advisor to review detailed aspects of any package proposal.

Equipment: the judicious use of equipment can certainly be a good way of exploring the potential of high technology and of adapting it for your needs in many cases, a relatively cheap and simple form of aquaculture can be improved significantly by the use of specific, often quite simple items of equipment-aerators in ponds, pumps to supplement water supply, screens or filters to remove contaminants from incoming water, simple mechanical lifts to assist harvesting, etc.

In other cases, eg tanks in hatcheries, cages for ongrowing the equipment represents the major part of the production system. In only a few cases in fact can equipment be defined as high-technology-eg electronic or computerised alarms or controls, water treatment systems, egg counting and sorting machines, etc-these are usually only used for specialised duties.

Probably the most important factors in using equipment successfully are:

- to be clear about what you want, what purpose the equipment has to serve, and in what conditions the equipment must operate.
- to choose equipment which is suited to the equipment of the farm, can be run by farm staff, and will withstand local operating conditions.
- to use reliable manufacturer, preferably one with local service capacity, and a proven record of use in conditions similar to yours.
- to be as clear as possible about the capital costs of the fully installed equipment, its operating and maintenance costs, its likely lifespan, and the probable effects on the yields running costs of the project as a whole.

- to install it properly, to use the correct operating procedures, and to monitor its use.

Normally, you will find quite a wide range of suppliers for the commonly used items, such as tanks nets, cages, feeders, etc. In addition many other items, such as pipes, valves, pumps, air blowers, air conditioners, forklifts, hoists etc. are supplied through standard outlets, ie not as specialised aquaculture equipment.

While the former supplier are often specialised in aquaculture, and develop their products specifically for the sector, they may lack the wider resources, and sometimes the technical backup, offered by larger general suppliers. On the other hand, these more general suppliers, unless specifically committed to the industry, may not be well enough acquainted with aquaculture and its requirements to offer the best technical guidance.

DEVELOPING VALUE

Increasingly, adding value into aquaculture products is recognised as a vital and necessary component of the production process. Whereas formerly producers could rely upon their chosen combination of production system technology to produce a sufficiently high value product it has become practice to attempt to add value wherever possible in the production process. Some explanation for his lies within the increased levels of competition within the international aquaculture sector. More rapid diffusion of technical innovations mean that few producers now can realistically expect to identify, develop and maintain a technically based comparative advantage for long. There is a constant need to innovate or at least adopt new technical standards.

To some extent the need to add value is dependent upon the existing value realised within the production process. In general it can be expected that more value added will tend to increase profitability.

But this in turn will tend to increase the number of players in the sector and hence production of that species is likely to rise and ultimately reduce market prices through the increased supply. This in itself then generates more pressure to add yet further value.

Adding value may be done in a variety of different ways, and not all require the incorporation of technology. Indeed it should be remembered that good husbandry practice is perhaps the most effective way in which value might be added. The increased speed of technical progress and its availability may mean that comparative advantages are more likely to be realised and be sustained through good husbandry practice.

The objective behind adding value should always be to the driving factor in any decision as to how any value might be added. Clearly there is little point in adding value unless this can be reflected in improved profitability. However this does necessarily mean that the addition of value will automatically bring about an increase in market prices. In some instances, especially those where competition is intense, adding value will serve simply to differentiate the output of one producer from that of another. Although, in the longer term it may prove possible to capitalise upon this product differentiation, to generate a larger market share.

In an increasingly cost conscious marketing environment the temptation to add value simply through increasing inputs, such as improved feeds, to the production process should also be avoided unless this is reflected in market gains. To compete within the wider market for foods, aquaculture products cannot afford the needless incorporation of additional inputs.

Perhaps the most important consideration determining the addition of value is whether the market will perceive, and itself place some value on, nay changes made. If this is not the case them the wisdom of so doing must be questioned; although longer strategic considerations also will need to be brought into the equation.

The whole notion of what constitutes perceived value must be recognised to be a dynamic phenomenon. Consumer preferences are in a state of permanent flux and are varied within individual market segments.

Thus in some target markets it maybe possible to add value by incorporating further processing of the finished product, for example by producing a fillet product rather than a whole fish. However other consumers may prefer that the fillets be available in a modified atmosphere pack, whilst others still might place greater value on a gutted head-on product. This demands that the producer remains in constant touch with the market to identify the nature and composition of any changes in customer wants. As was implicit in the above examples, it should also be noted that value may be perceived in the intangible features of the product. For example the packaging will additional benefits of convenience, easier storage etc. But increasingly so too may consumers perceive value in the inputs to their prospective purchase.

The environmental impact products is becoming a more important consideration within the purchase decision making process an cultured fish is an easily identifiable area for concern. Adding value may thus increasingly assume connotations of environmentally friendly production and other processes deemed to be consumer acceptable. However obscure, or ill-founded, these perceptions may appear to be, it must be remembered that they command the final dollar vote.

WHAT OF THE FUTURE?

Many of the more spectacular high-technology systems are still only under evaluation, in many cases a number of 'teaching problems' exist-unfortunately these often reflect the difference between what the promoters hoped could be done, and what the stock was prepared to accept! Indeed, systems such intensive recycle units have appeared, died out and reappeared again many times in the last twenty years-often with very little improvement during intervening period.

However, the aquaculture community is now at least more sceptical of wilder claims, more aware of the risks, and most importantly, more able to direct technical development towards the most useful and productive goals. We can therefore expect to see more substantial technical development in the future, based on more realistic, practical expectations. By using careful assessment, asking the right questions, and supporting and working with good and reliable suppliers, the aquaculturist can play an important part in making sure the industry has the technology it needs for the future.

SOCIO-ECONOMIC ASPECTS OF CHANGING FISH PRODUCTION SECTOR

Hungary is a landlocked country in Central Europe with an area of 93 000 square Km. The climate is typical continental climate with four distinct seasons. The fish growing season is about 150 days, when the temperature of the water is above 16°C. The average water temperature during the growing season is around 20°C. During the fairly long winter time from November until March, the ponds usually are freezing, and the over wintering of fish in deep ponds is needed. The annual precipitation is 800 mm, hut most of the rain falls out of the growing season. About 140 000 ha natural water area and 20 000 ha man made fish ponds are the major resources for fish production in Hungary. Due to the special hydrogeographic conditions, more than 90 percent of the rivers have their source in the surrounding countries, and this situation creates difficulties in the water management sometimes that has an effect on the fisheries and aquaculture as well. The geothermal water resources should also be mentioned as potential water resources for fish production.

81 freshwater fish species can be found in the natural waters in Hungary (Pintér, 1989), 68 specie are native in this region, while 13 species have been introduced mostly form Asia. There are 29 protected species in Hungary.

Fisheries in natural waters in one of the oldest activity in Hungary. For centuries fish "production" in river basins and wetlands were common. However, with the introduction of animal husbandry and the intensification of agricultural methods for crops, much of the lands were drained. In 19th century most of the rivers were "canalised" by flood protection dikes. Regulation of rivers and the drainage of swamps and marshes reduced the available fishing area and most of the natural spawning places disappeared. In order to compensate the decreasing captures from natural waters, extensive fish pond construction programme was started at the beginning of this century. Fish farms were built up on poor soils (not suitable for other agricultural activities).

Large size fish farms (up to 5000 hectares) have been dominant in pond fish production that were operated by large co-operatives and state farms under the centrally planned economy on the past forty-five years. The common carp dominant polyculture production is the most typical fish production method in Hungary, 94 percent of the total fish production were cyprinids in 1990, out of which 72 percent was common carp. There has been a steady increase in the fish production after the Second World War as shown in Figure 2. Although the vast majority of the fish was produced in state owned fish farms, that were operated according to the rules of the centrally planned economy, changes in market conditions have effected the production even in this period as illustrated in Figure 3. After a steady increase in the production between 1971 and 1977, there was a decline in the fish pond production in the following two years.

Besides the adverse climate conditions in this period, economical factors (suspension and stop of government subsidy) had a great effect on the decline of the production. Since this time the fish production sector has become market oriented, and the production was determined by the conditions of demand and supply on the market.

Since demand exceeded supply on the market the production increased again and reached an ever measured peak in 1983 with a total volume of 43,000 tons.

A considerable amount of the production were Chinese carps at this time, that were exported to the Middle East. Due to the war between Iraq and Iran, these markets collapsed suddenly and this was one of the main reasons why fish production dropped dramatically in the following years. Fish production showed a gradual increase between 1986 and 1988 but than there was a 10 percent drop in 1990 and 9 percent in 1991. The fish production was cut back by farmers because of the uncertainties in the ownership, and the increasing costs. The effect of the long drought period should also be mentioned, that affected adversely the natural production as well that has been declining since 1986.

Fish Production in Ponds

About 70 percent of the total production in Hungary comes from fish ponds. The area of the fish ponds used for fish production is 27,000 hectares, however only 20,700 hectares were in use in 1992. Common carp has been raised in ponds for many years in Hungary. At the beginning, the common carp was produced in monoculture. The traditional practice has been changed significantly in the late sixteen when the Chinese carps-silver carp (Hypophthalmichthys molitrix Val.), bighead (Aristichthys nobilis Rich.) and grass carp (Ctenopharyngodon idella Val.)-were introduced to Hungary. Since then, the fish production in polyculture has become dominant in the Hungarian fish farming, in which Chinese carps play an important role in increasing the production due to the full utilisation of food web. The market size of common carp is about 1 kg, and in normal conditions it takes three years to reach this size. At the end of the first growing season the average body weight of the fry if between 10 and 50 grams.

The fish is kept in special wintering ponds during the winter time and stocked into fingerling rearing ponds in the spring time when the water temperature increase above 10oC. The fish can reach an average body weight of about 150-250 grams after the second summer, and after a second wintering period it can be grown up to market size during the third summer. The fish production in ponds is based on the utilisation of the natural food, therefore the application of manure and fertilisers is a vital part of the technology, however supplementary feeding with cereals is also applied. 1.5-2.0 tons/ha total yield can be reached in a well managed fish ponds in a 150 days growing season. Strong seasonality exists in ponds fish production and in marketing: more than 70% of table size fish is harvested ans sold during the Christmas time.

Increased production costs, environmental impact problems and utilisation of poor quality soils were the main reasons for the development of integrated aquaculture in Hungary. Fish-cum-duck culture, aquaculture rotation and the use of organic manure of different sources in aquaculture systems were the typical forms applied. Significant extra production could be reached through these technologies. Among the more valuable cultured fishes the European catfish, or wells (Silurus glanis L.), pikeperch (Stizosteidon lucioperca L.,), pike (Esox lucius L.) and the eel (Anguilla anguilla L.) should be mentioned. For European catfish, pikeperch and pike special Hungarian methods of propagation and culture have been developed. Development of culture technologies for endemic crayfishes (Astacus astacus I, and Astacus leptodactylus L.) is a new direction in aquaculture.

Fish Production in Natural Waters

About 140,000 hectares natural water area is available in Hungary for aquaculture and fisheries. Approximately 30 percent of the total fish production comes from lakes (including the 60,000 hectares Lake Balaton), rivers, oxbow lakes. The applied management practices show a great variety according to bee the conditions of the water body.

The contribution of river fish ries to the total fish production is only 3 percent, but it has special importance to provided fresh fish to the market continuously and also to improve the selection of species on the market. Most of the oxbow lakes are utilised in a complex way in Hungary that includes flood production, water supply, run-off water collection, fisheries, duck production, recreation, and support of protected and endangered species).

The importance of recreation and recreational fisheries however has got more emphasis in the utilisation of these water areas. The relative importance of recreational fisheries has generally been increasing in Hungary due to the growing demand fort sport fishing. There are about 360,000 anglers in Hungary, and their contribution to the total market size fish production was as much as 24 percent in 1992.

Fish Production in Intensive Systems

The fish production in intensive systems is not more than about 500 tons annually, that is less than 2 percent of the total fish production in Hungary (Varadi and Tahy, 1993). There are eight intensive tank-production systems in Hungary, in which high-Valuc species, like sturgeons (Siberian sturgeon

(Acipenser baeri Brandt) and its hybrid with endemic sterlet (Acipenser ruthenus L.) eel (Anguila anguilla L.), European and African catfishes (Silurus glanis L. and Clarias gariepinus Burchel) are produced mostly for export. The operation of these systems is based on the utilisation of geothermal energy that is abundantly available in Hungary. These have been a steady but moderate increase in the intensive production of high-value species because of the good export marketing possibilities, availability of geothermal water resources and the changing economical environment, however the rapid expansion of these systems can not be expected in the near future because higher licensing criteria and stricter regulatory systems are going to be introduced.

SOCIO-ECONOMIC ASPECTS OF FISH CULTURE

The food and agriculture industry plays an important role in the Hungarian economy. It generates 17% of the GDP, employs 18% of the active population, supplies 25% of the total export, and contributes significantly to the maintenance of the positive balance of payment (Rasko, 1993). The fish production sector is a small, but special sub-sector of the agriculture. About 3,000 people are involved in fisheries and aquaculture currently, that is about 0.54 percent of the total labour force in aquaculture (Nash, 1992). The total value of fish production is about 23 million USD in the year of 1992 that remains below two percent of the total agricultural production. Fish does not play an important role in the diet of the people. The fish consumption per capita is not more than 3 kg/year, and the ratio of fish was about 4.0% of the total meat consumption (7403 kg/year/capita) in 1991.

Although fish is not a staple food in Hungary, it has a special, traditional role during certain holidays, first of all during Christmas time. Fish production is one of the most profitable agricultural activity, and there is an increasing interest by new entrepreneurs in fish production that can be carried out in the frame of small-scale private enterprises. Unfortunately no well established marketing channels are available, and fish processing practically doesn't exist in Hungary (most of he fish are sold alive). About 2,000 tons of fish mostly carp) was imported from the neighbouring Countries (Czech Republic, Croatia, Romania etc.) in 1992, on lower prices than the local production costs. At the same time there was a dramatic drop in export volume in 1992.

The member farms of the Association of Fish Producers exported 75 percent less market size carp in 1992 than in 1991. There was also a high competition on western markets with Eastern European producers, that were seeking for hard currency. Fish however has not only economical importance, but it has a special role in the optimal utilisation and conservation of the natural water resources in Hungary. Fish ponds are important water users, and about 50% of the total irrigation water is used for the water supply of fish ponds. Fish ponds however are not simply "objects" for fish production, but helps to retain rain and run-off waters, improves the micro climate and their aesthetic role should also be mentioned. The natural waters and their surroundings are becoming high value resources for the preservation of the native flora and fauna and also for recreation and eco-tourism. The fish is not only a user of natural waters as habitats, but indicates environmental problems and able to improve the conditions of degraded water bodies through biomanipulation. Recreational fisheries is one solid basis of he future development of the fish culture in Hungary, but the demand for fish as healthy food is also expected to be increased in the future. In spite of the rather insufficient advertisement, there is an increasing public awareness of the importance of healthy diet in which fish plays an important role. According to Kesteloot (1990) there is significant correlation between the animal fat consumption and the mortality of the population. The situation in Hungary is very bad in this respect, therefore the improvement of the diet should be an important task in the future. A new programme has been launched recently in order to improve the diet in Hungary, that includes the development of fish processing as well.

CHANGES OF THE FISH PRODUCTION SECTOR DURING THE TRANSITIONS PERIOD

In order to enabling Hungary to join the European integration, a rigorous and occasionally painful economic restructuring has been implemented. Before the restructuring, even in 1990, more than 80% of the arable lands was cultivated by large-scale co-operatives and state farms, and only 14% by small producers.

The proportion of private ownership has already reached 80%, and when the privatisation will be completed, only 2% of the arable land will stay permanently in state hands. A special element of the privatisation process is the compensation. According to the Compensation Law, the land (the area is maximised) is being returned to be original owners and/or their successors. Until the compensation process is completed, no state owned land can be sold, but the land can be rented (Rasko, 1993). In 1992 the former state owned co-operatives have been dissolved and "new type" of co-operatives were established, where the leaders were freely elected, the properties and assets went into the ownership of the members.

The law let the members to decide whether they want to stay in or leave the co-operative taking their share or assets with them. About 6-10% of the members decides to leave the co-operatives and to start individual enterprise. Those who left the co-operatives received about 5 ha land and some equipment with a value of about 4-5 thousand USD as an average. (Horvath et al., 1993). There are about 25 "new type" of co-operatives in Hungary that are involved in fish production. The activity of these co-operatives has not changed too much, however the ownership has been restructured and they have full autonomy.

Out of the 121 state farms only 25 remained in state ownership due to their strategic role in seed and breeding stock supply. Most of the state farms have been restructured and operate as share holding companies or limited liability companies, however ownership rights have been exercised by the State Property Agency for the time of being.

There have been 22 states owned fish farms in the past, and only two of them remained in state ownership. These two fish farms have special role in the Hungarian fish culture.

Balaton Fish Farm is responsible for the management of the 60,000 hectares lake Balaton that is a major tourist area in Hungary, the Warm Water Fish Hatchery at Szahalombatta, that was established by the help of FAO, is a basic hatchery for quality seed supply, and also acts a training and development centre. During the privatisation process, various size and type of fish farms have been and being established. Most of the private fish farms are fairly small in area, but there are private fish farms that operate several hundred ha fish ponds.

There are considerable changes in the legal, financial and institutional background as well, in order to facilitate the transition process into market economy. Numerous laws and acts have been passed in the past period such as: Compensation Law, Law on the Mandatory Restructuring of the State Owned enterprises, Law on Co-operative Transformation, Market Regime Law, to mention only the most important ones.

Although the new Fisheries Act has not been fully elaborated yet, a recently issued Ministerial Order brought some considerable changes in the application of the fisheries rights. According to this regulation, all companies, organisations and private entrepreneurs have equal right to apply for the fisheries right of a water area owned by the state. The Ministry gives priority to those applicants that have high professional skill and present a well elaborated plan for the fisheries utilisation of the given water body. In order to promote private enterprises various governmental interventions have been implemented in form of reduced taxes special loans and interest rates. A special Agricultural Fund has also been established recently to promote the restructuring process, to improve the technical conditions of agriculture production, and to preserve the gene stocks and natural resources.

One component of this special fund is the Fisheries Fund that provides financial support for the rehabilitation and fish enhancement of natural waters, and it also support research on this area. However other components of the Agricultural Fund can also be used by the fish production sector for the maintenance of high value fish breeding stocks and for the technical improvement of production facilities.

Financing is one of the main problems for fish farmers. Although the bank systems has been changing, special agricultural loans are not readily available, running times and interest rates are not tailored to the special conditions of the agricultural production. Difficulties in financing would create an even more serious liquidity crisis in fish production, where the production cycle is long, and it takes three years to grow the fish up to market size in traditional fish pond production. The respond of the farms to the increasing financial difficulties was the cutback of the production. In the centrally planned economy trade was regulated administratively rather than by the market. The producers were dependent on the various state owned trading companies that were in a monopolistic position having exclusive export-import licence.

Although foreign trade has been "liberalised" in the early eighties, business contracts and knowledge on this field remained the privilege of a fairly small number of experts. After being directed or guided by the state, most of the producers have not been prepared for the new challenges, they lack marketing skills, business and market information. Thanks in the entrepreneurial spirit of the Hungarian farmers, most of them are adapting well to the new situation. The new law on "Market Regime" helps to regulate the market and to secure free and fair competition for all participants on the market. According to this new law, so called "Product Councils" have been organised with the participation of the producers and traders. One branch of the product councils is the Fish council, that suggests limit prices, contingents, export-ban periods etc.

Under the previous regime, technology transfer was accomplished through the technical staff of agricultural scientist, economists, and veterinarians employed by the state farms and co-operatives, however this functional system has been disappeared during the restructuring of the agriculture. The reestablishment and strengthening of an effective extension system has been started, however it is not functional yet.

There are some firms and groups which provide private agricultural advisory services to farmers, however the main constraints confronting the establishment of a flourishing extension services by these new practitioners are given as: lack of money to pay for the service, in some areas community prejudice against former employees of the cooperatives and state farms, and farmer confidence in his ability to solve his own problems (Finlayson, 1993). Larger scale farming units such as many privatised state farms and larger co-operatives continue to employ their own technical advisory staff or purchase the required assistance at full cost from private extension services, universities or research institutions, but small scale farmers are still facing with difficulties to get access to special services and information. No extension service is available for fisheries and aquaculture. However the Fish Culture Research Institute and other institutions that are involved in research and development on this field, are providing assitance and services through various channels. The Association of Fish Producers also assists the technology transfer providing linkage between farmers and research institutions, organising meetings and training programmes. Agricultural education and training is undertaken by a number of institutions including three agricultural universities in Hungary. Although no MSc courses on fisheries are available at these universities, opportunities exist for students to be specialised in fisheries and aquaculture, an there are also state scholarships to study these subjects abroad. Vocational training and education is limited due to the relatively small demand by the sub-sector, and courses and post graduate education are held intermittently.

There is one special school in Hungary for the training of skilled fishermen. The regular 3-year course caters for 20-25 students annually, but post graduate master-fishermen courses are also held occasionally according to the demand. There is a solid research and development background in aquaculture and fisheries in Hungary. Special methods in the propagation and pond production of common carp that were elaborated by Hungarian scientists are widely applied world-wide.

The core institution for research and development is the Fish Culture Research Institute (HAKI) at Szarvas, that is an internationally acknowledged centre in this field. In spite of the serious cutback in financing research during the past three years, the Institute could preserve and even develop the valuable research capacities as a result of an internal restructuring programme.

CONCLUSIONS

The difficulties of the transition period into market economy combining with other difficulties related to the increasing competition for water resources and adverse weather conditions, created recession in fish production in the past four years.

The considerable changes in ownership created uncertainty in the fish production sector. As a result of this, not only the production was cut back, but the necessary reconstruction works of he ponds and water control structures were also neglected. Even in money was available for some farmers, they used if for new investments rather than reconstruct facilities that might not be his own in the future. Most of the farmers are not prepared for the new challenges, business and market information are not readily available, the extension background has not been developed. The fish production sector should meet the requirements of an emerging more diversified market that requires continuous supply of high quality products, and the new regulations of the European Community requires special measures for producers and processors.

Financing is a serious bottle neck of the development of fish production. The financial institutional and bank system has not been fully developed yet, agricultural loans are not readily available and the interest rate is very high. Water is becoming a critical factor in fish production due to the increasing competition for water resources.

The difficulties in water supply have further been increased by the drought in the past ten years. Due to environmental considerations, the use of natural waters for aquaculture is becoming limited, while the importance of recreational fisheries is increasing. During the further development of aquaculture, farmers have to be prepared for implementing more effective measures in order to minimise the pollutants in the effluents since higher licensing criteria and stricter regulatory system will be applied in the near future.

Bibliography

Ausubel, F.M.: *Current Protocols in Molecular Biology*, New York, John Wiley and Sons, 1989.

Barnum, Susan R.: *Biotechnology: An Introduction*, Belmont, Thomson/Brooks/Cole, 2005.

Broach, J.R.: *The Molecular Biology of the Yeast*, Cold Spring Harbor, Cold Spring Harbor Laboratory, 1981.

Chrispeels, Maarten : *Plants, Genes and Crop Biotechnology*, Sudbury MA, Jones and Barlett Publishers, 2003.

Curtis, H.: *Biology*, New York, Worth Publishers, 1989.

Dixon, Dougal: *After Man-A Zoology of the Future*, New York, St. Martin's Press, 1981.

Fransman M, Junne G, Roobeek A: *The Biotechnology Revolution?*, Oxford, Blackwell, 1995.

Fumento, Michael: *Bioevolution: How Biotechnology is Changing Our World*, San Francisco, Encounter Books, 2003.

Goodman, D.C.: *From Farming to Biotechnology: A Theory of Agro-industrial Development*, Oxford, Blackwell, 1987.

Moravec, Hans: *Mind Children: The Future of Robot and Human Intelligence*, Cambridge, Harvard University Press, 1988.

Murray, David: *Seeds of Concern: The Genetic Manipulation of Plants*, Sydney, University of New South Wales, 2003.

Primrose, S.B.: *Principles of Gene Manipulation*, London, Blackwell Scientific Publications, 1989.

Shetty, Kalidas: *Food Biotechnology*, New York, Dekker/CRC Press, 2005.

Smith, John E.: *Biotechnology*, Cambridge, Cambridge University Press, 2004.

Towle, Albert: *Modern Biology*, Austin, Holt, Rinehart and Winston, 1988.

INDEX

□□□